Essentials of Physical and Theoretical Chemistry: Unveiling the Molecular Universe

Welcome to a captivating journey through the intricate realms of molecules, energy, and the fundamental forces that shape the very fabric of our world. In the pages that follow, we embark on an exploration that delves deep into the heart of matter, revealing the hidden landscapes of atoms and molecules, and uncovering the underlying principles that govern their behavior.

"Essentials of Physical and Theoretical Chemistry: Unveiling the Molecular Universe" is an odyssey that takes us beyond the visible, delving into the unseen forces that dictate chemical reactions, molecular interactions, and the dynamics of matter. This book is a guided tour through the fascinating world where science and imagination converge to illuminate the complexities of our physical reality.

As we delve into the chapters, we'll encounter the foundational concepts that lay the groundwork for physical and theoretical chemistry. From the quantization of energy levels and the wave-particle duality of matter to the laws of thermodynamics and quantum mechanics, we'll journey through the theories that have reshaped our understanding of the universe at its most fundamental levels.

But this is more than a theoretical discourse. As we explore the molecular universe, we'll encounter the tools and techniques that empower us to experimentally probe the behaviors of matter. We'll witness the power of spectroscopy as it peels back the layers of molecular structure, and we'll appreciate the art of mathematical models that predict the behavior of particles on both the macroscopic and quantum scales.

"Unveiling the Molecular Universe" is more than just a catchphrase; it encapsulates the essence of this exploration. We will unravel the mysteries behind chemical reactions, the intricacies of molecular motion, and the driving forces behind the formation of matter as we know it. The book doesn't just seek to provide answers; it encourages questions, contemplation, and a desire to dive further into the intricacies of the molecular realm.

Whether you're a student embarking on a journey of discovery, a seasoned researcher seeking to expand your horizons, or a curious mind intrigued by the forces that shape our world, this book invites you to peer through the lens of physical and theoretical chemistry. Together, we'll navigate the uncharted territories of molecular behavior, challenge our perceptions, and unlock a deeper understanding of the wonders that surround us.

So, let us begin this adventure – a voyage that promises to enrich our intellect, spark our curiosity, and illuminate the mysteries of the molecular universe that is both our foundation and our destiny.

Introduction

- The Significance of Physical and Theoretical Chemistry
- The Quest to Unveil the Molecular Universe

Chapter 1: Foundations of Matter and Energy
- Particle-Wave Duality: Unifying Nature's Dual Faces
- Energy Quantization: Unraveling the Atomic Spectrum
- The Uncertainty Principle: Navigating Quantum Limits

The Significance of Physical and Theoretical Chemistry

Physical and theoretical chemistry lie at the heart of our understanding of the natural world, playing a pivotal role in unraveling the complexities of matter and energy. These branches of chemistry delve into the fundamental principles that govern the behavior of molecules, atoms, and particles on both macroscopic and quantum scales. Their significance extends far beyond the laboratory walls, permeating virtually every facet of science and technology. Here, we explore the profound significance of physical and theoretical chemistry in shaping our understanding of the molecular universe and driving innovation across various disciplines.

1. Understanding Molecular Behavior: Physical and theoretical chemistry offer insights into the intricate behaviors of molecules, elucidating phenomena such as molecular interactions, reaction kinetics, and energy transfer. By studying the forces that bind atoms together, scientists can predict and explain chemical reactions, allowing for the design of new materials, drugs, and catalysts. This understanding forms the foundation of the chemical sciences, enabling scientists to harness the power of matter to create transformative technologies.

2. Predicting and Modeling Reactions: Through theoretical calculations and quantum mechanical simulations, physical and theoretical chemistry provide the tools to predict and model complex chemical reactions. These predictions aid researchers in identifying reaction pathways, understanding reaction mechanisms, and optimizing reaction conditions. This ability to foresee the outcomes of chemical processes accelerates the development of environmentally friendly and energy-efficient processes, revolutionizing industries from pharmaceuticals to energy production.

3. Bridging the Microscopic and Macroscopic: Physical and theoretical chemistry act as a bridge between the microscopic world of atoms and molecules and the macroscopic world that we perceive. By elucidating the connections between the behavior of individual particles and the properties of bulk materials, these disciplines enable us to comprehend the emergence of macroscopic phenomena from the interactions of countless tiny components.

4. Advancing Materials Science: The understanding of molecular properties and interactions provided by physical and theoretical chemistry drives advancements in materials science. From designing novel materials with specific properties to engineering materials for various applications, such as electronic devices, polymers, and nanomaterials, these disciplines underpin the development of materials that shape our modern world.

5. Innovations in Energy Research: As the world seeks sustainable energy solutions, physical and theoretical chemistry play a critical role in advancing energy research. From designing more efficient batteries and solar cells to exploring new catalysts for clean energy conversion, these disciplines contribute to the development of technologies that address global energy challenges.

6. Inspiring Cross-Disciplinary Collaboration: The principles of physical and theoretical chemistry transcend traditional disciplinary boundaries. They foster collaborations between chemists, physicists, biologists, engineers, and beyond. This cross-disciplinary approach enables researchers to tackle complex problems from multiple perspectives, driving innovation and expanding the frontiers of knowledge.

7. Driving Quantum Technological Advances: In the age of quantum technologies, physical and theoretical chemistry are central to unlocking the potential of quantum computing, quantum communication, and quantum sensors. These technologies promise to revolutionize computation, communication, and measurement by harnessing the principles of quantum mechanics.

In essence, physical and theoretical chemistry provide the intellectual foundation upon which modern science and technology rest. They empower us to understand the underlying principles that govern the behavior of matter, enabling us to manipulate and harness these principles for a better future. From unraveling the mysteries of the smallest particles to propelling innovation in diverse fields, physical and theoretical chemistry remain at the forefront of scientific exploration and technological advancement.

The Quest to Unveil the Molecular Universe

In the vast expanse of the cosmos, our journey to understand the universe begins with an exploration of the smallest constituents of matter – molecules. The quest to unveil the molecular universe is a relentless pursuit that transcends the boundaries of the observable, delving into the intricate dance of atoms and particles that shape the very fabric of reality. This journey, guided by the principles of physical and theoretical chemistry, takes us from the mundane to the extraordinary, from the familiar to the mysterious, as we peel back the layers of existence to reveal the fundamental truths that govern our world.

1. Journey into the Unseen: The molecular universe is a realm where the naked eye falls short, where forces beyond our immediate perception govern the behavior of matter. With the aid of spectroscopy, we harness the power of light to unravel the stories hidden within the molecular spectrum. From the absorption of photons to the resonance of nuclei, each interaction whispers secrets that narrate the tales of molecular structure, motion, and transformation.

2. Unraveling the Quantum Riddle: As we delve deeper, we encounter the enigma of quantum mechanics. In this realm, particles transcend traditional notions of reality, existing in states of uncertainty, bound by probabilities and wave functions. The quest to unravel this riddle leads us to the Schrödinger equation, a mathematical gateway to a universe where particles dance in the realms of probability, and where the very act of observation influences their nature.

3. Beyond the Microscopic: Physical and theoretical chemistry empower us to bridge the gap between the microscopic and macroscopic worlds. The principles we uncover at the atomic and molecular scale reverberate through the tapestry of reality, shaping everything from the properties of materials to the intricacies of chemical reactions. We glimpse the interplay between forces and energy, between entropy and order, and we realize that the molecular universe orchestrates the symphony of life itself.

4. Architects of Creation: In this quest, we become architects of creation. Armed with the principles of quantum chemistry, we predict molecular properties, simulate reaction mechanisms, and engineer new materials. We wield computational tools that offer us a glimpse into the future, allowing us to design molecules with precision, and harness their potential for transformative technologies.

5. Inspiring Innovations: Our journey through the molecular universe extends beyond the confines of the laboratory. Physical and theoretical chemistry inspire cross-disciplinary collaborations that lead to innovations across fields. From quantum technologies to sustainable energy solutions, from nanomaterials to drug discovery, our pursuit of understanding at the molecular level drives advances that shape the world we inhabit.

6. Embracing the Unknown: As we venture further, we encounter the uncharted frontiers of the molecular universe. Emerging technologies promise new insights, from harnessing the power of quantum computing to manipulating matter at the nanoscale. With each step, we embrace the unknown, eager to uncover the mysteries that lie beyond the horizons of our current knowledge.

The quest to unveil the molecular universe is a journey of wonder, curiosity, and discovery. It challenges us to explore the fundamental nature of reality, to peer into the inner workings of matter, and to unveil the profound interconnectedness that weaves through the fabric of existence. It is a journey that beckons us to transcend the limits of our perception, inviting us to embark on an odyssey that redefines our understanding of the universe – from the subatomic to the cosmic, from the ordinary to the extraordinary.

Particle-Wave Duality: Unifying Nature's Dual Faces

In the pursuit of understanding the nature of matter and energy, one of the most profound and counterintuitive revelations emerged: the concept of particle-wave duality. This principle, at the heart of quantum mechanics, challenges our conventional understanding of the universe by revealing the dual nature of particles and their inherent wave-like characteristics. Particle-wave duality stands as a cornerstone of modern physics, bridging the realms of classical mechanics and quantum theory, and ultimately unifying nature's dual faces in a breathtaking symphony of waves and particles.

1. The Quantum Paradox: In the classical world, particles were conceived as discrete entities with well-defined positions and velocities. However, as scientists delved deeper into the microscopic realm, they encountered puzzling phenomena that defied classical explanations. The double-slit experiment, for instance, unveiled that particles like electrons and photons could exhibit interference patterns, much like waves do. This apparent contradiction challenged the very foundations of our understanding.

2. The Wave-Particle Duality Revelation: The breakthrough came when scientists realized that particles, at the subatomic level, could display both particle-like and wave-like behaviors. This revelation led to the formulation of the wave-particle duality principle. According to this principle, particles possess both discrete localized properties (particle behavior) and spread-out, wave-like characteristics (wave behavior) simultaneously.

3. The Mathematics of Waves and Particles: The mathematics of wave-particle duality is encapsulated in the concept of the wave function. This mathematical expression represents the probability distribution of a particle's position and momentum. It elegantly captures the duality, portraying particles as propagating waves of probability that interfere constructively or destructively, leading to the observed patterns in experiments.

4. The Uncertainty Principle: Wave-particle duality is intimately tied to Heisenberg's Uncertainty Principle. This principle posits that the more accurately we know a particle's position, the less accurately we can know its momentum, and vice versa. This intrinsic uncertainty, a fundamental consequence of wave-like behavior, underpins the probabilistic nature of quantum mechanics.

5. Applications and Implications: Particle-wave duality is not merely an abstract concept; it has profound practical implications. It plays a central role in understanding the behavior of electrons in atoms, the interactions of particles in quantum mechanics, and the functioning of electronic devices. Moreover, it forms the basis of technologies such as electron microscopy and quantum computing.

6. A New Paradigm: Particle-wave duality challenges our intuition and compels us to expand our understanding of reality. It paints a picture of the universe where the distinction between particles and waves blurs, revealing a universe where both aspects are intertwined. In this new paradigm, particles become waves of probability, and the very act of measurement collapses these probabilities into definite outcomes.

7. The Unity of Nature: In its essence, particle-wave duality unifies the seemingly disparate worlds of classical and quantum physics. It reminds us that nature, at its core, defies easy categorization and embraces a tapestry of behaviors that transcend our ordinary perceptions. As we unravel the mysteries of particle-wave duality, we inch closer to unifying nature's dual faces, recognizing that waves and particles are not opposing entities but harmonious expressions of the underlying fabric of reality.

Energy Quantization: Unraveling the Atomic Spectrum

In the realm of quantum mechanics, the concept of energy quantization stands as a fundamental pillar that reshaped our understanding of matter and its interactions with light. Through this principle, scientists unearthed the intricate tapestry of the atomic spectrum, revealing that energy levels within atoms are not continuous but discrete, like rungs on a ladder. Energy quantization not only explained the enigmatic patterns observed in spectroscopy but also laid the groundwork for understanding atomic structure, chemical bonding, and the fascinating dance of electrons within atoms.

1. The Birth of Energy Quantization: The classical view of energy envisioned a continuous spectrum where particles could possess any energy value. However, the pioneering work of Max Planck in the early 20th century shattered this notion. Planck introduced the concept of quantization to explain the distribution of energy in black-body radiation, marking the birth of quantum theory.

2. Quantum States and Energy Levels: Energy quantization reveals that within an atom, energy is restricted to specific levels, known as quantum states. These energy levels are analogous to the rungs of a ladder, where electrons can exist only on these discrete steps and not in between. This insight explains the characteristic patterns observed in atomic spectra.

3. The Atomic Spectrum: Energy quantization finds its most striking manifestation in the atomic spectrum. When atoms absorb or emit light, they do so in quantized packets called photons. The energy of these photons corresponds to the energy difference between the allowed energy levels within the atom. This gives rise to the unique lines and bands observed in spectroscopy, offering a distinctive fingerprint for each element.

4. Balmer, Lyman, and Beyond: One of the most celebrated achievements of energy quantization is the explanation of spectral lines in hydrogen, known as the Balmer series. This series, and subsequent series like the Lyman and Paschen series, unveiled the intricate relationships between energy levels and provided insights into the structure of the atom.

5. Quantum Jumps and Emission Spectra: Energy quantization introduces the concept of quantum jumps, where electrons transition between discrete energy levels by absorbing or emitting photons. This process accounts for the sharp lines in emission spectra and underscores the discrete nature of energy in the atomic realm.

6. The Bohr Model and Beyond: Niels Bohr's revolutionary model of the hydrogen atom incorporated energy quantization, providing a framework for understanding atomic structure and explaining the spectral lines. While Bohr's model has limitations, it marked a significant step in linking quantized energy levels to the behavior of electrons within atoms.

7. Implications and Modern Insights: Energy quantization extends beyond the atomic spectrum. It underpins our understanding of chemical bonding, explaining why molecules possess specific electronic configurations. In the quantum mechanical realm, it shapes our comprehension of electron behavior, offering insights into molecular shapes and properties.

8. Energy Quantization's Legacy: Energy quantization revolutionized our understanding of energy and matter. It served as a cornerstone in the development of quantum mechanics and transformed our perception of atomic and molecular phenomena.

Its legacy endures in every facet of modern physics, from spectroscopy and electronics to the exploration of nanoscale materials and beyond, reminding us that the continuous fabric of reality is woven from discrete threads of energy.

The Uncertainty Principle: Navigating Quantum Limits

In the realm of quantum mechanics, where particles dance in a probabilistic ballet, the Uncertainty Principle emerges as a profound revelation. Coined by Werner Heisenberg, this principle unveils a fundamental limitation inherent to the nature of particles and waves. It dictates that the more precisely we know a particle's position, the less precisely we can know its momentum, and vice versa. The Uncertainty Principle navigates the intricate boundary between certainty and ambiguity, transforming our understanding of the microscopic world and challenging our notions of measurement and reality.

1. The Precision Predicament: In classical mechanics, the precision of measurements was limited only by the accuracy of our instruments. However, as we descend into the quantum realm, the Uncertainty Principle introduces a level of inherent uncertainty that transcends any technological advancement. It asserts that certain pairs of properties, like position and momentum, cannot be precisely determined simultaneously.

2. Heisenberg's Insight: The Uncertainty Principle was formulated by Werner Heisenberg in 1927. He realized that particles, behaving both as waves and particles, cannot have perfectly defined properties. The more accurately we know a particle's position, the less accurately we can know its momentum, and vice versa.

3. Quantum Limits: The Uncertainty Principle sets a lower limit to the product of the uncertainties in position and momentum. This means that as we strive for higher precision in measuring one property, the uncertainty in the other property must increase. This inherent trade-off arises from the wave-like nature of particles, where their position and momentum become entangled in a delicate dance.

4. Implications for Measurement: The Uncertainty Principle introduces a fundamental ambiguity into the act of measurement. It tells us that particles don't have well-defined properties until we measure them. This challenges our classical intuition, where we often assume that particles possess inherent properties regardless of measurement.

5. Quantum Reality and Observations: The Uncertainty Principle raises profound questions about the nature of reality and the role of observation. It implies that particles only acquire well-defined properties upon measurement. This notion has led to philosophical debates about the observer's impact on the observed, blurring the lines between objective reality and subjective observation.

6. Applications and Quantum Mechanics: The Uncertainty Principle is not a mere theoretical concept; it has practical implications. It underlies the stability of atoms and the formation of chemical bonds. Moreover, it's a foundational principle in quantum mechanics, guiding our understanding of quantum states, wave functions, and the probabilistic nature of particle behavior.

7. Beyond Heisenberg: While Heisenberg's formulation remains a cornerstone, other formulations and interpretations of the Uncertainty Principle have emerged, deepening our grasp of its implications. From the mathematical framework of quantum mechanics to the entanglement of particles, the principle continues to spark exploration and inquiry.

8. Embracing Quantum Uncertainty: The Uncertainty Principle compels us to embrace the inherent limits of knowledge at the quantum level. It challenges our desire for certainty and certainty, inviting us to confront the paradoxes that arise from the interplay of particles and waves. In this intricate dance of uncertainty, we uncover the essence of the quantum world, where ambiguity and possibility converge in a captivating symphony of probabilities.

Schrödinger Equation: The Heartbeat of Quantum Mechanics

At the core of quantum mechanics beats a mathematical symphony that orchestrates the behavior of particles in the quantum realm—the Schrödinger equation. Crafted by Austrian physicist Erwin Schrödinger in 1925, this equation encapsulates the essence of quantum theory, describing how the wave functions of particles evolve over time. It is the fundamental tool that unveils the dance of electrons, the stability of atoms, and the intricate play of probabilities that governs the quantum world.

1. The Birth of a Quantum Revolution: As classical physics proved inadequate to describe the behavior of particles at the atomic scale, the quest for a new theory led to the birth of quantum mechanics. Schrödinger's formulation emerged as a response to the dual nature of particles, harmonizing the wave-like behavior of matter with the mathematics of waves.

2. The Quantum State Equation: The Schrödinger equation is a differential equation that links the time evolution of a quantum state (often described as a wave function) to the energy of a system. It encodes the dynamics of particles by describing how their wave functions change with time and interact with their surroundings.

3. Wave Function and Probability: The wave function is the heart of quantum mechanics. It is a mathematical representation that encapsulates the probability amplitude of finding a particle at a specific position and time. The square of the wave function's magnitude gives the probability density of finding the particle in a particular state.

4. Operators and Observables: Schrödinger's equation introduces operators that act on the wave function, representing physical observables such as position, momentum, and energy. These operators yield measurable quantities when applied to the wave function, allowing us to predict the outcomes of measurements.

5. Quantum Mechanics in Action: Solving the Schrödinger equation unveils the allowed energy states and wave functions of a system. From electrons orbiting nuclei to complex molecules, the equation provides insights into the behavior of particles within various contexts, shaping our understanding of chemical bonding, spectroscopy, and material properties.

6. Boundaries and Approximations: Solving the Schrödinger equation is not always straightforward due to its mathematical complexity. Different approximations and techniques are employed to address various systems. These approximations, like the Born-Oppenheimer approximation for molecules, enable us to explore quantum phenomena across scales.

7. Wave-Particle Duality Revisited: The Schrödinger equation bridges the gap between the wave-like and particle-like behavior of particles. It shows how particles' wave functions evolve over time, including scenarios where particles can exist in superpositions—a concept that challenges our classical intuition.

8. Beyond Single Particles: Extensions of the Schrödinger equation lead to quantum field theories, exploring particle interactions and relativistic effects. Additionally, it plays a central role in understanding entanglement, a phenomenon where particles become correlated in ways that defy classical explanations.

9. A Quantum Odyssey: The Schrödinger equation isn't just a mathematical tool; it's the essence of quantum reality. It guides our exploration of the quantum universe, from unraveling the mysteries of subatomic particles to shaping the technologies of tomorrow. It reminds us that at the heart of the quantum realm, waves and particles engage in a dance choreographed by a fundamental equation, revealing the extraordinary complexity hidden within the fabric of reality.

Quantum Numbers: Mapping Electron Configurations

In the intricate world of quantum mechanics, where electrons dance in probabilistic orbits, quantum numbers emerge as crucial coordinates that map out the complex landscape of atomic and molecular behavior. These numbers provide a framework for understanding the distribution of electrons in atoms and molecules, guiding us through the quantum labyrinth and revealing the unique fingerprints of chemical elements. From delineating energy levels to describing orbital shapes, quantum numbers are the compasses that navigate the quantum world.

1. Principal Quantum Number (n): The principal quantum number, symbolized by 'n', designates the main energy level or shell in which an electron resides. It quantifies the size and energy of an orbital, with higher values of 'n' corresponding to higher energy levels and larger orbitals.

2. Angular Momentum Quantum Number (l): The angular momentum quantum number, denoted by 'l', defines the shape of the orbital within a given energy level. It takes on values from 0 to (n-1), indicating different subshells. For instance, 'l=0' corresponds to an s orbital, 'l=1' to a p orbital, and so on.

3. Magnetic Quantum Number (m_l): The magnetic quantum number, represented as 'm_l', specifies the orientation of the orbital in space relative to a coordinate axis. It takes on values ranging from -l to +l, determining the spatial distribution of electron density within the orbital.

4. Spin Quantum Number (m_s): The spin quantum number, denoted as 'm_s', describes the intrinsic spin of an electron. It can have two values: +½ (spin up) or -½ (spin down). This quantum number accounts for the electron's magnetic behavior and is a fundamental component of the Pauli exclusion principle.

5. Pauli Exclusion Principle: Quantum numbers underpin the Pauli exclusion principle, a fundamental rule that states that no two electrons in an atom can have the same set of quantum numbers. This principle forms the basis for the organization of electrons within atoms and the arrangement of electrons in different orbitals.

6. Electron Configurations: Quantum numbers pave the way for determining electron configurations—the arrangement of electrons in various energy levels, subshells, and orbitals. Electron configurations provide insights into an element's chemical reactivity, as well as its stability and properties.

7. Orbital Filling Rules: Quantum numbers guide the order in which electrons fill orbitals within an atom. The Aufbau principle dictates that electrons occupy the lowest energy orbitals first. Hund's rule suggests that within a subshell, electrons tend to occupy separate orbitals before pairing up, minimizing repulsion.

8. Spectroscopic Notation: Quantum numbers find their place in spectroscopic notation, a concise representation of an element's electron configuration. This notation uses the principal quantum number followed by the letter designating the subshell, and superscripts indicating the number of electrons.

9. The Quantum World Unveiled: Quantum numbers reveal the multifaceted world of atomic and molecular behavior, guiding us through the maze of energy levels, orbitals, and electron distributions. They provide a mathematical framework that underpins the intricate properties of matter and form the cornerstone of quantum mechanics—a realm where electrons

dance to the tune of quantum numbers, creating the symphony of the chemical elements.

The Quantum Mechanical Model: A Glimpse Inside the Atom

In the early 20th century, the quest to understand the inner workings of the atom led to a paradigm shift in our conception of reality—the birth of the quantum mechanical model. This model, grounded in the principles of quantum mechanics, revolutionized our understanding of atomic structure, revealing a complex world where particles exhibit both particle-like and wave-like behaviors. With its intricate equations, probabilities, and uncertainty, the quantum mechanical model offered a tantalizing glimpse into the enigmatic realm within the atom.

1. A Crisis of Classical Physics: Classical physics failed to explain the intricacies of atomic behavior. The collapse of the classical model, which depicted electrons orbiting the nucleus in fixed paths, necessitated a new approach to understand atomic phenomena.

2. Enter Quantum Mechanics: Quantum mechanics emerged as the solution—a theory that marries the wave-like behavior of particles with mathematical equations that predict probabilities. This new framework, born from the works of Planck, Einstein, Heisenberg, Schrödinger, and others, shattered classical determinism.

3. Probability Clouds and Wave Functions: The quantum mechanical model employs wave functions to describe the probability distribution of an electron's position within an atom. These wave functions form probability clouds that reveal the likelihood of finding an electron in a particular region of space.

4. Electron Orbitals: Unlike classical orbits, where electrons follow defined paths, the quantum mechanical model introduces the concept of electron orbitals. These are three-dimensional regions around the nucleus where the probability of finding an electron is highest. Orbitals are characterized by their energy, shape, and orientation.

5. Quantum Numbers Revisited: Quantum numbers, as introduced by the quantum mechanical model, describe the quantized energy levels, orbital shapes, orientations, and spins of electrons. They offer a blueprint for understanding electron distribution and behavior.

6. Wave-Particle Duality in Action: The quantum mechanical model showcases wave-particle duality—an electron's behavior as both a wave and a particle. This duality manifests in phenomena such as diffraction and interference, offering insights into the behavior of electrons and particles.

7. Uncertainty Principle and Quantum Limits: The Uncertainty Principle, a cornerstone of quantum mechanics, influences the quantum mechanical model. It imposes limits on the precision with which we can know certain pairs of properties, fundamentally shaping the behavior of particles and our understanding of reality.

8. Quantum Numbers and Electron Configurations: Quantum numbers play a crucial role in the quantum mechanical model, mapping out the arrangement of electrons in atoms. They guide electron configurations, helping us understand chemical properties, reactivity, and the periodic table.

9. From Atoms to Molecules: The quantum mechanical model extends beyond atoms to molecular structures. It guides our understanding of chemical bonding, molecular shapes, and spectroscopic properties, laying the foundation for the study of complex compounds.

10. Bridging the Microscopic and Macroscopic: The quantum mechanical model bridges the gap between the microscopic and macroscopic worlds. It underpins the behavior of matter at the atomic scale and influences the properties of materials on a larger scale, revolutionizing fields such as materials science and nanotechnology.

The quantum mechanical model stands as a testament to the intricate beauty of the quantum realm. It challenges our classical intuitions, offering a new lens through which to glimpse the mesmerizing dance of particles and waves within the atom. As we navigate its complex equations and probabilistic landscapes, we embark on a journey into the heart of the quantum world —a realm that continues to astonish, inspire, and reshape our understanding of the universe.

Molecular Orbitals: Weaving Electron Clouds

In the captivating world of molecular chemistry, where atoms unite to form intricate compounds, the concept of molecular orbitals emerges as a fundamental framework that weaves together the behavior of electrons within molecules. Molecular orbitals are the ethereal landscapes where electrons move, dance, and interact, shaping the properties and reactivity of compounds. Through this intricate dance of electron clouds, molecular orbitals illuminate the essence of chemical bonding, molecular structure, and the diverse tapestry of molecules that populate our world.

1. Atomic Orbitals and Molecular Formation: Before diving into molecular orbitals, it's crucial to revisit atomic orbitals—regions around atomic nuclei where electrons reside. When atoms combine to form molecules, these atomic orbitals combine, overlap, and interact to create molecular orbitals.

2. Molecular Orbital Theory: Molecular orbital theory, grounded in quantum mechanics, introduces the concept of molecular orbitals—regions around a molecule where electrons are likely to be found. These orbitals arise from the linear combination of atomic orbitals, both bonding and antibonding.

3. Bonding Molecular Orbitals: When atomic orbitals overlap in phase, they create bonding molecular orbitals. Electrons residing in these orbitals stabilize the molecule by sharing their density between the atomic nuclei, forming covalent bonds that hold atoms together.

4. Antibonding Molecular Orbitals: Conversely, when atomic orbitals overlap out of phase, they form antibonding molecular orbitals. Electrons in these orbitals have higher energy and disrupt the stability of the molecule by reducing electron density between the nuclei.

5. Sigma and Pi Bonds: Molecular orbitals give rise to distinct types of bonds. Sigma (σ) bonds form head-on overlap of atomic orbitals, while pi (π) bonds arise from sideways overlap. Double and triple bonds, essential in molecular diversity, result from combinations of these bond types.

6. Electron Density and Shape: Molecular orbitals define the electron density distribution within a molecule. The shape of the electron density cloud provides insights into molecular geometry, impacting properties like polarity and reactivity.

7. Molecular Orbitals and Spectroscopy: Molecular orbitals have a profound impact on spectroscopic properties. Electronic transitions between these orbitals lead to absorption and emission of light, giving rise to UV-Vis, IR, and fluorescence spectroscopy—a window into molecular structure and behavior.

8. Hybridization and Orbital Hybridization: Orbital hybridization transforms atomic orbitals into hybrid orbitals, allowing for more accurate descriptions of molecular shapes and bond angles. Hybridization plays a pivotal role in understanding molecular geometry and properties.

9. Insights into Reactivity and Bonding: Molecular orbitals provide insights into chemical reactivity, explaining the selectivity of chemical reactions and the stability of compounds. They guide our understanding of why some molecules bond strongly while others remain inert.

10. Quantum Mechanical Symphony: Molecular orbitals are the quantum mechanical symphony orchestrating the intricate ballet of electrons within molecules. They form the bedrock of molecular chemistry, unraveling the secrets of bonding, structure, and properties that underpin the diversity of compounds that compose our world.

As we delve into the realm of molecular orbitals, we become explorers of the electron's ethereal dance—the poetry of electron clouds shaping the molecules that define our existence. From simple diatomic molecules to complex macromolecules, the concept of molecular orbitals is a key to unlocking the hidden melodies of chemical diversity and the harmonious interplay of electrons within the intricate tapestry of molecules.

Bonding and Antibonding: Constructive and Destructive Overlaps

In the captivating world of molecular chemistry, the concepts of bonding and antibonding orbitals play a pivotal role in shaping the stability, properties, and reactivity of molecules. These two types of molecular orbitals arise from the constructive and destructive interference of atomic orbitals as atoms come together to form compounds. Through their intricate dance, bonding and antibonding orbitals illuminate the delicate balance between attraction and repulsion, creating the foundation for our understanding of chemical bonding and molecular structure.

1. Constructive Overlaps: Bonding Orbitals: When atomic orbitals of neighboring atoms overlap in phase, they create bonding orbitals. This overlap is akin to a harmonious collaboration, where electrons are shared between nuclei. Bonding orbitals are regions of high electron density between the atoms, fostering the covalent bonds that hold molecules together.

2. Destructive Overlaps: Antibonding Orbitals: On the flip side, when atomic orbitals overlap out of phase, they form antibonding orbitals. This interaction results in destructive interference, causing reduced electron density between the atoms. Electrons in antibonding orbitals disrupt the cohesive forces between atoms, weakening the bond.

3. Formation of Molecular Orbitals: The interplay of bonding and antibonding orbitals is a dynamic process that emerges when atomic orbitals combine. In the formation of molecular orbitals, the constructive overlap of atomic orbitals leads to bonding orbitals, while the destructive overlap gives rise to antibonding orbitals.

4. Sigma (σ) and Pi (π) Bonds: The concept of bonding and antibonding orbitals plays a crucial role in understanding different types of chemical bonds. Sigma (σ) bonds form from the head-on overlap of atomic orbitals, while pi (π) bonds arise from the sideways overlap of p orbitals. Multiple bonds, such as double and triple bonds, involve the participation of both types of bonds.

5. Stability and Energy Levels: Bonding orbitals lower the overall energy of the molecule, contributing to its stability. Electrons occupying these orbitals facilitate the sharing of electron density between atoms. Antibonding orbitals, on the other hand, raise the energy of the molecule and discourage electron sharing, weakening the bond.

6. Molecular Orbital Diagrams: Molecular orbital diagrams provide a visual representation of the arrangement of bonding and antibonding orbitals. The energy levels of these orbitals determine whether a molecule is stable or reactive and shed light on its potential chemical behavior.

7. Impact on Molecular Properties: The presence of bonding and antibonding orbitals profoundly influences molecular properties such as bond strength, bond length, and molecular stability. The interplay between these orbitals dictates whether a molecule will readily react or remain inert.

8. Beyond Individual Bonds: The concept of bonding and antibonding extends beyond individual bonds to the understanding of molecular systems as a whole. The collective effect of multiple bonds and interactions shapes the overall stability and behavior of complex molecules.

Bonding and antibonding orbitals represent the intricate choreography of electrons within molecules—the push and pull that define the delicate balance between attraction and repulsion. They guide the formation and breaking of bonds, shaping the identity of chemical compounds and the countless reactions that drive the processes of life and matter. Through their dance, we gain insights into the very essence of molecular existence, unraveling the mysteries of molecular structure, reactivity, and the diverse tapestry of the chemical world.

Hybridization: Adapting to Molecular Structures

In the intricate world of molecular chemistry, the concept of hybridization emerges as a versatile tool that enables atoms to adapt their atomic orbitals and form unique molecular structures. Hybridization plays a pivotal role in shaping the geometry, reactivity, and properties of molecules, allowing atoms to optimize their electron arrangements for a wide range of bonding situations. Through this remarkable process, hybridization transforms ordinary atomic orbitals into hybrid orbitals, paving the way for a deeper understanding of molecular shapes and the diversity of compounds that populate our chemical universe.

1. The Need for Adaptation: Atoms seldom possess the exact atomic orbitals needed to form certain types of bonds or achieve specific molecular geometries. Hybridization arises as a response to this challenge, allowing atoms to adapt their electron configurations to create stable and diverse molecules.

2. Hybrid Orbitals: A New Blueprint: Hybridization transforms atomic orbitals into hybrid orbitals—composite orbitals with characteristics of multiple atomic orbitals. This process involves mixing different types of atomic orbitals, leading to new hybrid orbitals that are better suited for bonding.

3. sp Hybridization: Linear Geometry: In sp hybridization, one s orbital and one p orbital combine to form two sp hybrid orbitals. This hybridization results in linear geometry, as seen in molecules like acetylene (C_2H_2). These hybrid orbitals facilitate the formation of sigma (σ) bonds.

4. sp² Hybridization: Trigonal Planar Geometry: sp² hybridization involves the mixing of one s orbital and two p orbitals, creating three sp² hybrid orbitals. This hybridization leads to trigonal planar geometry, as observed in molecules like ethene (C_2H_4). These hybrid orbitals participate in both sigma (σ) and pi (π) bonds.

5. sp³ Hybridization: Tetrahedral Geometry: In sp³ hybridization, one s orbital and three p orbitals combine to form four sp³ hybrid orbitals. These orbitals result in tetrahedral geometry, as seen in molecules like methane (CH_4). sp³ hybridization accounts for sigma (σ) bonds in tetrahedral structures.

6. sp³d Hybridization: Trigonal Bipyramidal Geometry: Hybridization can extend to d orbitals, as in sp³d hybridization. This process forms five sp³d hybrid orbitals, resulting in trigonal bipyramidal geometry as seen in phosphorus pentafluoride (PF_5).

7. sp³d² Hybridization: Octahedral Geometry: Further hybridization, as in sp³d² hybridization, involves six hybrid orbitals and leads to octahedral geometry as seen in sulfur hexafluoride (SF_6).

8. Insights into Molecular Geometry: Hybridization offers insights into molecular geometry, impacting bond angles and shapes. It explains why certain molecules adopt specific geometries and how hybrid orbitals enable atoms to optimize their electron distribution.

9. Beyond Geometry: Impact on Reactivity: Hybridization influences molecular reactivity. It affects the availability of orbitals for bonding and the stability of intermediates in chemical reactions, shedding light on reaction mechanisms and products.

10. Tailoring Electron Arrangements: Hybridization exemplifies the adaptability of atoms in forming stable compounds. By altering their electron configurations, atoms can tailor their properties to suit a wide range of bonding situations, showcasing the remarkable versatility of chemistry.

Through the process of hybridization, atoms transcend the confines of their native atomic orbitals, forging new paths in the creation of compounds. This transformative process highlights the intricate balance between adaptation and stability, unveiling the elegance with which nature shapes the world of molecules. Hybridization is the toolkit that empowers atoms to embrace versatility and create the diverse molecular landscape that defines the tapestry of chemistry.

Laws of Thermodynamics: Principles of Energy Conservation

In the realm of physical sciences, the laws of thermodynamics stand as pillars of understanding that govern the behavior of energy and matter. These principles provide a roadmap for comprehending the intricate dance of heat, work, and energy flow in various systems. From the conservation of energy to the limitations of efficiency, the laws of thermodynamics unveil the fundamental truths that shape our understanding of the physical world and drive our technological advancements.

1. First Law of Thermodynamics: Conservation of Energy: The first law, often referred to as the law of conservation of energy, states that energy cannot be created or destroyed; it can only be transferred or transformed from one form to another. This law underpins the principle that the total energy of an isolated system remains constant over time.

2. Energy Transfer and Work: The first law encompasses both the transfer of energy in the form of heat and the performance of work. When energy is transferred to a system as heat or work, the internal energy of the system increases. Conversely, when energy is transferred out of the system, the internal energy decreases.

3. Internal Energy and Heat: The internal energy of a system is the sum of its kinetic and potential energies at the molecular level. Heat refers to the energy transferred between systems due to temperature differences. The first law quantifies how heat and work contribute to changes in the internal energy of a system.

4. Second Law of Thermodynamics: Entropy and Disorder: The second law introduces the concept of entropy—a measure of the amount of energy in a system that is unavailable for doing work. It states that in any energy transfer or transformation, the total entropy of a closed system always increases over time, leading to the concept of increasing disorder or randomness.

5. Entropy and Energy Quality: The second law highlights the concept of energy quality. Energy spontaneously flows from higher-quality forms (e.g., high-temperature heat) to lower-quality forms (e.g., low-temperature heat), with a corresponding increase in entropy. This law places limitations on the efficiency of energy conversion processes.

6. Carnot Efficiency and Thermodynamic Processes: The Carnot efficiency, based on the second law, sets an upper limit to the efficiency of heat engines. It defines the maximum possible efficiency that can be achieved by a heat engine operating between two temperature reservoirs. Thermodynamic processes, such as isothermal, adiabatic, and isobaric, arise from the interplay of heat and work in various systems.

7. Third Law of Thermodynamics: Absolute Zero and Entropy: The third law states that as the temperature of a system approaches absolute zero (0 K), the entropy of the system approaches a minimum value. Absolute zero represents the lowest possible temperature, where the motion of particles nearly ceases. This law has implications for understanding the behavior of matter at extremely low temperatures.

8. Applications in Science and Engineering: The laws of thermodynamics find applications in various fields, from engineering and physics to chemistry and biology. They provide insights into energy efficiency, heat transfer, refrigeration, phase transitions, and the behavior of systems at different energy states.

9. The Thermodynamic Universe: The laws of thermodynamics weave a comprehensive tapestry that shapes our understanding of energy conservation, entropy, and the intricate workings of the universe. They are fundamental to grasping the behavior of matter and energy and have far-reaching implications for technological advancements and scientific discoveries.

The laws of thermodynamics, standing as pillars of scientific understanding, empower us to decipher the energy dynamics of the universe and unravel the mysteries of energy conservation, entropy, and order. They serve as beacons of knowledge, guiding our exploration of the physical world and enabling us to harness the forces of nature for the betterment of our society.

Entropy and Enthalpy: The Tug-of-War for Disorder and Energy

In the captivating realm of thermodynamics, the concepts of entropy and enthalpy engage in a constant tug-of-war that shapes the behavior of matter and energy within systems. These fundamental principles offer insights into the forces that drive reactions, transformations, and the delicate balance between order and disorder. As entropy pushes for increased randomness and enthalpy seeks to maintain or release energy, the interplay of these concepts provides a nuanced understanding of energy dynamics and the dynamics of chemical and physical processes.

1. Entropy: A Measure of Disorder: Entropy is a measure of the randomness or disorder of a system. The second law of thermodynamics states that the entropy of an isolated system tends to increase over time, leading to greater disorder. As molecules move and interact in myriad ways, they contribute to the overall increase in entropy.

2. Entropy and Energy Distribution: The concept of entropy can be likened to the spreading out of energy within a system. As energy spreads out, the number of possible microstates—different ways that particles can arrange themselves—increases, leading to higher entropy.

3. Entropy in Phase Transitions: Phase transitions, such as melting, vaporization, and sublimation, provide practical examples of entropy changes. When a solid transitions to a liquid or gas, the increase in possible arrangements of particles results in higher entropy.

4. Enthalpy: The Heat of the Matter: Enthalpy, often symbolized as 'H', represents the total heat content of a system at constant pressure. It includes internal energy, pressure-volume work, and energy exchanged as heat. Changes in enthalpy (ΔH) reveal the heat absorbed or released in a process.

5. Endothermic and Exothermic Reactions: In chemical reactions, the enthalpy change indicates whether a reaction is endothermic (absorbing heat) or exothermic (releasing heat). An endothermic reaction absorbs energy from the surroundings, while an exothermic reaction releases energy to the surroundings.

6. Gibbs Free Energy: The Driving Force: The relationship between entropy and enthalpy is encapsulated in the concept of Gibbs free energy (G). Gibbs free energy combines the effects of enthalpy and entropy changes to determine whether a reaction is spontaneous or not. A negative ΔG indicates a spontaneous reaction.

7. Balance between Entropy and Enthalpy: In chemical processes, there is a constant interplay between entropy and enthalpy. While entropy may drive systems towards greater randomness, enthalpy can oppose this by providing energy constraints. The balance between these factors determines whether a reaction will occur spontaneously.

8. Entropy and Life: Entropy's influence extends to biological systems. Organisms maintain their internal order by utilizing energy from their surroundings. The increase in entropy outside the organism is offset by the decrease in entropy within it.

9. Practical Implications: Energy and Efficiency: Entropy and enthalpy have practical implications for energy efficiency and engineering. They guide our understanding of energy transfer, heat engines, and refrigeration systems.

10. The Dance of Disorder and Energy: Entropy and enthalpy engage in an intricate dance that governs the fate of matter and energy. As entropy seeks to increase randomness and enthalpy contributes to energy considerations, their interplay shapes the course of chemical reactions, phase transitions, and the behavior of systems in the universe.

Entropy and enthalpy—the twin forces that dictate the balance between chaos and energy—are the essence of the dynamic world of thermodynamics. Through their interactions, we uncover the mechanisms that drive change, the energy exchanges that power our world, and the delicate equilibrium that governs the universe's dance of matter and energy.

Gibbs Free Energy: The Thermodynamic Driving Force

In the realm of thermodynamics, where energy and matter interplay to shape the behavior of systems, the concept of Gibbs free energy stands as a guiding beacon—a thermodynamic driving force that determines the spontaneity and direction of chemical reactions. Through its intricate equation and insightful interpretations, Gibbs free energy unveils the underlying forces that govern whether a reaction will proceed spontaneously, providing a crucial tool for understanding the dynamics of the physical and chemical world.

1. Defining Spontaneity: Gibbs free energy, symbolized as 'G', is a thermodynamic potential that quantifies the energy available to do useful work in a system at constant temperature and pressure. A key tenet is that nature tends to move towards lower energy states, seeking stability and equilibrium.

2. Gibbs Free Energy Equation: The Gibbs free energy change (ΔG) for a process is calculated using the equation: $\Delta G = \Delta H - T\Delta S$, where ΔH is the change in enthalpy, ΔS is the change in entropy, and T is the absolute temperature. The sign and magnitude of ΔG reveal whether a process is spontaneous, nonspontaneous, or at equilibrium.

3. Driving Forces: Spontaneity and Equilibrium: A negative ΔG indicates a spontaneous process—a reaction that can proceed without continuous external intervention. If ΔG is positive, the process is nonspontaneous and requires energy input. When ΔG is zero, the system is at equilibrium, and there is no net change.

4. Entropy and Enthalpy Contributions: The equation $\Delta G = \Delta H - T\Delta S$ underscores the balance between the enthalpy change (ΔH) and the temperature-dependent entropy change ($T\Delta S$). Enthalpy contributes to the heat exchanged in a reaction, while entropy accounts for the dispersal of energy.

5. Enthalpy-Entropy Compensation: In some cases, changes in enthalpy and entropy may compensate for each other, resulting in near-constant ΔG values over a range of temperatures. This phenomenon highlights the intricate relationship between energy and disorder.

6. Chemical Equilibrium: A ΔG Perspective: For a reaction at equilibrium, ΔG is zero, indicating that the forward and reverse reactions occur at the same rate. Deviations from equilibrium ($\Delta G \neq 0$) drive reactions towards attaining equilibrium and the lowest possible free energy state.

7. Reaction Favorability and ΔG Magnitude: A more negative ΔG indicates a more favorable reaction, with greater spontaneity. The magnitude of ΔG also indicates the extent to which a reaction will proceed, providing insights into the equilibrium composition of reactants and products.

8. ΔG and Nonstandard Conditions: For reactions occurring under nonstandard conditions, the standard Gibbs free energy change ($\Delta G°$) is adjusted using the equation $\Delta G = \Delta G° + RT \ln(Q)$, where R is the gas constant and Q is the reaction quotient.

9. Practical Applications: Thermodynamics in Action: Gibbs free energy guides our understanding of phase transitions, chemical reactions, and equilibria. It plays a pivotal role in various fields, including chemistry, biology, materials science, and engineering.

10. Energy, Entropy, and Direction: Gibbs free energy stands as a compass in the complex landscape of energy and entropy. It directs us towards a deeper understanding of why reactions proceed, offering insights into the energetics that drive matter's transformative journey through the universe.

Gibbs free energy—the compass of spontaneity—is a cornerstone of thermodynamics, empowering us to navigate the complex terrain of energy and matter interactions. Through its equation and principles, we decipher the codes that guide the universe's journey towards equilibrium, providing us with the tools to harness energy, predict reactions, and uncover the mysteries of the natural world.

Reaction Rates: Investigating Reaction Dynamics

In the dynamic world of chemistry, the concept of reaction rates serves as a window into the intricate dance of molecules as they transform and rearrange. Reaction rates offer insights into the speed at which chemical reactions occur, shedding light on the pathways, mechanisms, and factors that influence the progression of reactions. Through careful analysis of reaction rates, scientists unlock the secrets of reaction dynamics, enabling us to design and optimize processes, understand natural phenomena, and manipulate molecular behavior.

1. The Quest for Speed: Reaction Rates Defined: Reaction rate is the measure of how quickly reactants transform into products in a chemical reaction. It quantifies the change in concentration of reactants or products over time and provides a tangible metric for the speed of reactions.

2. Rate Laws and Reaction Order: Rate laws describe the relationship between the reaction rate and the concentrations of reactants. They reveal the order of the reaction with respect to each reactant, showcasing how changes in concentration influence the overall reaction rate.

3. Differential Rate Laws: Unraveling Reaction Orders: Differential rate laws express the reaction rate as a function of concentration. The exponents in these equations represent the reaction orders. The sum of the exponents for all reactants constitutes the overall reaction order.

4. Integrated Rate Laws: Time and Concentration: Integrated rate laws express concentration as a function of time. They provide insights into how the concentration of reactants changes as the reaction proceeds. Different integrated rate laws correspond to different reaction orders.

5. Half-Life: Time and Transformation: The half-life of a reaction is the time it takes for the concentration of a reactant to decrease by half. Half-lives provide a practical way to compare the relative speeds of different reactions and gauge their progress.

6. Factors Influencing Reaction Rates: Collision Theory: Collision theory proposes that reactions occur when reactant molecules collide with sufficient energy and proper orientation. Factors such as temperature, concentration, surface area, and catalysts influence the frequency and effectiveness of collisions, thereby impacting reaction rates.

7. Reaction Mechanisms: The How and Why: Reaction mechanisms elucidate the step-by-step pathways by which reactions occur. They involve a series of elementary steps that collectively lead to the overall reaction. Reaction intermediates and rate-determining steps play pivotal roles in these mechanisms.

8. Transition State Theory: Activation Energy and Transition States: Transition state theory delves into the concept of transition states—the fleeting configurations that molecules adopt during the course of a reaction. Activation energy represents the energy barrier that reactant molecules must overcome to reach the transition state and proceed to product formation.

9. Catalysts: Accelerating Reaction Paths: Catalysts are substances that increase the rate of a reaction without being consumed themselves. They provide alternative reaction pathways with lower activation energies, facilitating faster reaction kinetics.

10. Kinetics and Beyond: Insights and Applications: Reaction kinetics—the study of reaction rates—provides insights into the fundamental nature of chemical reactions. This understanding underpins diverse applications in fields such as industrial processes, environmental science, pharmacology, and materials engineering.

11. Unveiling Molecular Transformations: Reaction rates serve as investigative tools that unveil the intricate transformations of matter at the molecular level. They grant us access to the dynamic world of chemical reactions, empowering us to comprehend the intricacies of molecular motion and rearrangement that shape our physical and chemical universe.

Through the lens of reaction rates, we peer into the heart of chemical dynamics—the rapid and transformative dance of atoms and molecules. This journey unravels the secrets of reaction mechanisms, kinetic barriers, and the subtle interplay of factors that orchestrate the symphony of molecular change. From understanding everyday reactions to deciphering complex processes, the study of reaction rates empowers us to manipulate matter and harness the power of chemistry to mold the world around us.

Rate Laws and Reaction Orders: Unraveling Rate-Determining Steps

In the intricate tapestry of chemical kinetics, the concepts of rate laws and reaction orders illuminate the pathways and intricacies of reaction dynamics. These fundamental principles provide a blueprint for understanding how the concentrations of reactants influence the speed of chemical reactions. By dissecting the reaction rates and determining the reaction orders, scientists uncover the rate-determining steps that guide the progression of reactions, offering insights into the underlying molecular mechanisms and opening doors to control, optimization, and discovery.

1. Rate Laws: Mapping Concentration and Speed: Rate laws express the mathematical relationship between the rate of a reaction and the concentrations of its reactants. They provide a quantitative means of describing how changes in reactant concentrations impact the speed at which products form.

2. Reaction Orders: Exponents of Influence: Reaction orders define the power to which the concentration of a reactant is raised in the rate law expression. These exponents indicate the sensitivity of the reaction rate to changes in concentration. The sum of the reaction orders gives the overall reaction order.

3. Zero-Order Reactions: Independent of Concentration: In zero-order reactions, the rate is independent of the concentration of the reactants. The reaction rate remains constant, regardless of the changes in reactant concentrations. Such reactions often involve processes with limited reactant availability or surface reactions.

4. First-Order Reactions: Linear Concentration Dependence: First-order reactions display a linear dependence of the reaction rate on the concentration of a single reactant. The reaction rate is directly proportional to the concentration, leading to exponential decay in reactant concentration over time.

5. Second-Order Reactions: The Power of Doubling: Second-order reactions exhibit a quadratic dependence on reactant concentration or involve the collision of two reactant molecules. Doubling the concentration of a reactant results in a fourfold increase in the reaction rate.

6. Reaction Half-Lives and Order: Unveiling Time Scales: The reaction order influences the relationship between the reaction rate and the concentration of reactants. This relationship impacts the time it takes for reactant concentrations to decrease to half their initial values—a concept fundamental to understanding reaction dynamics.

7. Integrated Rate Laws: Time and Transformation: Integrated rate laws relate the concentration of reactants to time, offering insights into the time-dependent behavior of reactions. Different reaction orders yield distinct mathematical expressions for integrated rate laws.

8. Rate-Determining Steps and Mechanisms: Reaction orders and rate laws provide clues about the sequence of events in a reaction mechanism. The slowest step, known as the rate-determining step, governs the overall reaction rate. Understanding this step unveils the critical processes that dictate reaction kinetics.

9. Beyond Simple Reactions: Complex Kinetics: While simple reactions often exhibit well-defined rate orders, complex reactions involving multiple steps may have varying reaction orders depending on the reactants' roles in each step. In such cases, reaction mechanisms become crucial to unraveling kinetics.

10. Control and Optimization: Harnessing Kinetics: The insight gained from rate laws and reaction orders empowers scientists and engineers to control and optimize chemical processes. By manipulating reactant concentrations or employing catalysts, reactions can be accelerated or tuned to achieve desired outcomes.

Through the lens of rate laws and reaction orders, the complex dance of molecules comes into focus. These concepts decode the language of reaction kinetics, revealing the intricate interplay of concentrations, reaction rates, and molecular events. As we unravel the mysteries of rate-determining steps, we uncover the secret mechanisms that govern the pace of reactions, forging pathways to innovation, efficiency, and a deeper understanding of the molecular world.

Catalysis: Expediting Reactions with Molecular Partners

In the realm of chemistry, where reactions dictate the transformations of matter, the art of catalysis emerges as a powerful strategy to accelerate and control these processes. Catalysis involves the use of catalysts—molecular partners that facilitate reactions without being consumed themselves. By lowering activation energies and providing alternative reaction pathways, catalysts unlock new avenues for chemical transformations, offering benefits that span from industrial processes to environmental preservation and pharmaceutical discoveries.

1. Catalysts: Agents of Transformation: Catalysts are substances that enhance the rates of chemical reactions by providing an alternative reaction pathway with lower activation energy. They enable reactions to occur under milder conditions and increase the yield of desired products.

2. Homogeneous Catalysis: Integrating with Reactants: In homogeneous catalysis, the catalyst is in the same phase as the reactants. Catalysts form temporary complexes with reactants, facilitating the breaking and forming of bonds. Transition metal complexes are often employed in this type of catalysis.

3. Heterogeneous Catalysis: Interfaces of Reaction: Heterogeneous catalysis involves catalysts in a different phase from the reactants. Solid catalysts interact with reactants at the catalyst's surface, providing a platform for reactions to occur. Examples include catalytic converters in automobiles and industrial processes like ammonia synthesis.

4. Enzymes: Nature's Catalysts: Enzymes are biological catalysts that play essential roles in living organisms. They accelerate reactions in a highly specific and efficient manner, enabling vital biochemical processes to occur within narrow temperature and pH ranges.

5. Activation Energy: Catalysts as Energy Saviors: Catalysts lower the activation energy—the energy barrier that reactant molecules must overcome to reach the transition state and proceed to product formation. By providing an alternative, lower-energy reaction pathway, catalysts make reactions more accessible.

6. Reaction Pathways: Catalysts' Guiding Hands: Catalysts influence reaction pathways by altering the sequence of intermediates and transition states. They may facilitate the formation of key intermediates or stabilize transition states, thereby increasing reaction rates.

7. Mechanisms and Catalytic Cycles: Insights into Catalysts' Roles: Understanding catalytic mechanisms and cycles is crucial for designing effective catalysts. Catalysts participate in a series of reactions, and their regeneration enables them to continue facilitating multiple rounds of the same reaction.

8. Catalytic Promoters and Inhibitors: Fine-Tuning Reactivity: Promoters enhance catalytic activity, often by modifying the catalyst's surface properties or coordination environment. Inhibitors, on the other hand, reduce activity. These substances allow for precise control over reactions.

9. Industrial and Environmental Applications: Driving Progress: Catalysis drives numerous industrial processes, including petroleum refining, polymer production, and sustainable energy generation. It also plays a critical role in mitigating environmental pollution and developing green technologies.

10. Pharmaceuticals and Beyond: Catalysts in Discovery: Catalysis has transformative implications in drug development. It enables the creation of specific chiral molecules, which are crucial in pharmaceuticals. Catalysts also contribute to green chemistry initiatives and sustainable materials production.

Catalysis—this artful partnership between molecules—transcends mere reaction acceleration; it empowers us to harness the inherent potential of matter and control its transformations. As catalysts pave the way for cleaner, more efficient processes, they revolutionize industries, preserve ecosystems, and illuminate the pathways to innovative discoveries. In the realm of catalysis, molecules collaborate as architects of progress, and through their dynamic interactions, they shape a world of accelerated potential and unlimited possibilities.

UV-Vis Spectroscopy: Peering into Electronic Transitions

In the captivating world of spectroscopy, UV-Vis (Ultraviolet-Visible) spectroscopy stands as a versatile and insightful technique that allows us to peer into the electronic transitions of molecules. By probing the absorption and transmission of light within the ultraviolet and visible regions of the electromagnetic spectrum, UV-Vis spectroscopy unveils the energy shifts that occur as electrons transition between different electronic states. This technique not only provides a window into molecular structure and bonding but also enables the analysis of chromophores, complexation, and the quantification of analytes —a journey that spans from fundamental research to diverse applications across fields.

1. **Absorption of Light: Electronic Excitations:** UV-Vis spectroscopy focuses on the absorption of light by molecules due to electronic transitions. These transitions involve the promotion of electrons from lower-energy (ground) states to higher-energy (excited) states, often leading to color changes in the molecule.

2. **Spectrophotometers: Instruments of Insight:** Spectrophotometers, the workhorses of UV-Vis spectroscopy, measure the intensity of light absorbed or transmitted by a sample. By comparing the intensity of incident and transmitted light, the absorption spectrum of a sample is generated.

3. **Beer-Lambert Law: Quantitative Insights:** The Beer-Lambert law relates the concentration of a sample to the amount of light

absorbed. This fundamental relationship enables the quantitative determination of analytes and their concentration in solutions.

4. Absorption Spectra: Fingerprinting Molecules: Absorption spectra—the graphical representations of absorption intensity versus wavelength—provide distinctive patterns for different molecules. The position and intensity of absorption bands offer clues about the electronic transitions and the chromophores present.

5. Chromophores: The Absorption Architects: Chromophores are groups of atoms responsible for the absorption of light in a molecule. They contain π-electron systems that undergo electronic transitions, resulting in characteristic absorption bands.

6. Auxochromes: Shaping Absorption Spectra: Auxochromes are functional groups that modify the intensity and position of absorption bands. They enhance the color of chromophores and enable the fine-tuning of absorption properties.

7. Solvent Effects: A Molecular Soliloquy: Solvents play a pivotal role in UV-Vis spectroscopy. They can influence electronic transitions by altering the polarity and environment of molecules, leading to shifts in absorption bands.

8. Applications Across Fields: A Spectrum of Utility: UV-Vis spectroscopy finds applications in diverse fields, from chemistry and biology to environmental science and material characterization. It enables the analysis of dyes, monitoring enzymatic reactions, studying reaction kinetics, and quantifying concentrations.

9. Limitations and Complementary Techniques: The Full Picture: While UV-Vis spectroscopy provides valuable insights, it has limitations in distinguishing complex mixtures and elucidating detailed structural information. Complementary techniques, such as IR and NMR spectroscopy, fill these gaps.

10. Unveiling Molecular Stories: UV-Vis spectroscopy offers a narrative of electronic transitions—the tales of electrons moving between energy states. This narrative informs us about molecular structure, electronic properties, and the essence of chromophores, painting a vibrant portrait of molecules in the canvas of the electromagnetic spectrum.

UV-Vis spectroscopy, with its ability to unravel the stories of electronic transitions, takes us on a journey through the quantum world of molecules. From unraveling the secrets of chromophores to quantifying analytes and exploring the intricacies of molecular behavior, this technique illuminates the scientific landscape with its insights and applications. Whether in the laboratory or in the realm of innovation, UV-Vis spectroscopy guides our exploration of matter's electronic intricacies, bringing the hidden electronic tales of molecules into the light.

Infrared Spectroscopy: Vibrational Fingerprinting

Within the realm of spectroscopy, infrared (IR) spectroscopy emerges as a powerful technique that unveils the vibrational fingerprints of molecules. By probing the absorption of infrared radiation, IR spectroscopy offers a window into the intricate vibrations of atoms within a molecule. These vibrations reflect the unique composition and connectivity of atoms, enabling us to identify functional groups, elucidate molecular structures, and delve into the complex interactions that govern molecular behavior—a journey that spans from the foundational principles of molecular vibrations to a myriad of practical applications across scientific disciplines.

1. Vibrational Modes: Dancing Atoms and Bonds: Vibrational modes involve the oscillation of atoms and bonds within a molecule. Stretching and bending vibrations occur as bonds elongate and contract, or atoms move in specific patterns. Each functional group contributes distinct vibrational modes.

2. Infrared Radiation: Vibrational Resonance: Infrared radiation possesses energy levels that match the vibrational energies of molecules. When infrared light is absorbed, it excites molecules to higher vibrational states, creating a spectrum that captures the unique vibrational fingerprint of the molecule.

3. Spectral Ranges: A Multitude of Information: IR spectroscopy covers different spectral ranges: the near-IR, mid-IR, and far-IR. The mid-IR region, where most organic molecules absorb, is particularly informative for structural elucidation.

4. IR Absorption Bands: Peaks of Information: IR spectra display absorption bands, each corresponding to a specific vibrational mode. The position, intensity, and shape of these bands provide critical information about the types of bonds and functional groups present.

5. Functional Group Identification: A Spectral Puzzle: IR spectroscopy enables the identification of functional groups within a molecule. Peaks corresponding to characteristic vibrations—such as C=O, N-H, and O-H stretches—serve as diagnostic markers.

6. Hydrogen Bonding and Intermolecular Interactions: Hydrogen bonding and other intermolecular interactions can significantly influence IR spectra. They lead to shifts in absorption bands, offering insights into molecular associations and bonding strengths.

7. Quantitative Analysis: Concentration and Spectra: IR spectroscopy finds application in quantitative analysis. The Beer-Lambert law relates absorbance to concentration, enabling the quantification of analytes in solutions.

8. Solid-State IR Spectroscopy: Investigating Solids: Solid-state IR spectroscopy provides information about crystal structures, polymorphism, and the arrangement of molecules within solid materials.

9. Attenuated Total Reflectance (ATR): Probing Surfaces: ATR is a technique that enables the analysis of solid and liquid samples in their native form without extensive sample preparation. It is particularly useful for studying surfaces and coatings.

10. Applications Across Fields: Illuminating Molecular Behavior: IR spectroscopy is invaluable in chemistry, materials science, biology, and beyond. It aids in drug development, forensic analysis, environmental monitoring, and the study of biomolecular structures.

IR spectroscopy—this journey through vibrational landscapes—connects us to the harmonic symphony of atoms and bonds within molecules. It uncovers the secrets of functional groups, unveils the nuances of molecular interactions, and empowers us to decipher the intricate language of molecules. From elucidating molecular structures to exploring dynamic processes, IR spectroscopy paints a vivid picture of molecular behavior, bridging the gap between the quantum realm and our macroscopic world.

NMR Spectroscopy: Navigating the Nuclear Magnetic Resonance

In the captivating world of spectroscopy, Nuclear Magnetic Resonance (NMR) emerges as a magnetic compass that navigates the intricate landscapes of molecules. By exploiting the magnetic properties of atomic nuclei, NMR spectroscopy offers a detailed map of molecular structures, revealing the connectivity, environments, and dynamics of atoms within compounds. This technique deciphers the language of nuclear spins and chemical shifts, opening pathways to unraveling molecular mysteries, studying biomolecules, and advancing fields from chemistry to medicine—a journey that spans from the principles of nuclear magnetism to transformative applications across scientific disciplines.

1. Nuclear Spins: Quantum Dance of Nuclei: Nuclear spins give rise to magnetic moments, creating a tiny compass within atomic nuclei. These spins interact with external magnetic fields and provide the foundation for NMR spectroscopy.

2. Resonance Phenomenon: Resonating with Nuclei: NMR exploits the phenomenon of nuclear resonance, where nuclei absorb and emit radiofrequency radiation in the presence of a strong magnetic field. This interaction reveals unique information about atomic environments.

3. Chemical Shifts: Elemental Cartography: Chemical shifts encode the magnetic environment of nuclei and are measured in parts per million (ppm) relative to a reference compound. They reflect the electron distribution around the nucleus, aiding in identifying functional groups.

4. Spin-Spin Coupling: Vicinal Magnetic Dialogues: Spin-spin coupling arises from the magnetic interactions between adjacent nuclei. These couplings lead to splitting of NMR peaks and provide insights into the connectivity and arrangement of atoms within molecules.

5. NMR Spectrometers: Magnetic Observatories: NMR spectrometers are the instruments that facilitate NMR experiments. They generate powerful magnetic fields, manipulate radiofrequency radiation, and detect the signals emitted by nuclei.

6. Proton NMR: Unveiling Proton Environments: Proton NMR is a widely used technique that explores the environments of hydrogen nuclei in molecules. It offers insights into chemical shifts, coupling patterns, and molecular structure.

7. Carbon NMR: Mapping Carbon Skeletons: Carbon NMR reveals information about carbon environments within molecules. It helps determine the types of carbon atoms present and their connectivity.

8. Multidimensional NMR: Unraveling Complex Structures: Multidimensional NMR techniques provide high-resolution insights into complex molecules and biomolecules. They enhance spectral resolution by introducing additional dimensions of information.

9. NMR in Biomolecular Studies: Biomolecular Dialogues: NMR plays a pivotal role in studying biomolecules such as proteins and nucleic acids. It reveals dynamic processes, protein-ligand interactions, and the three-dimensional structures of biomolecules.

10. Applications Across Fields: From Molecules to Medicine: NMR spectroscopy has diverse applications, from determining molecular structures and studying reaction mechanisms to drug discovery and medical imaging (MRI).

NMR spectroscopy—this magnetic conversation with the atomic world—transports us into the realm of quantum spins and molecular dialogues. It uncovers the hidden stories of chemical shifts, spin couplings, and nuclear environments, enabling us to decipher molecular mysteries with unparalleled precision. From the laboratory to the clinic, NMR spectroscopy enriches our understanding of matter, revealing the secrets of molecules and their behavior, and shaping the landscape of scientific discovery.

Boltzmann's Distribution: Predicting Particle Behavior

In the realm of statistical mechanics, Boltzmann's distribution stands as a foundational principle that unveils the intricate dance of particles within a system. Derived from the marriage of classical mechanics and probability theory, this distribution offers insights into the distribution of energy among particles in different energy states. By predicting the probabilities of particles occupying various states, Boltzmann's distribution provides a window into the behavior of systems at thermal equilibrium, shedding light on phenomena ranging from molecular motion to the behavior of gases—a journey that spans from the kinetic theory of gases to its wide-ranging applications across scientific disciplines.

1. Kinetic Theory of Gases: A Molecular Ballet: Boltzmann's distribution originates from the kinetic theory of gases, which models gases as ensembles of rapidly moving particles. It links the microscopic motion of particles to macroscopic observables like pressure and temperature.

2. Energy States and Probability: Mapping the Landscape: In a system with different energy states, particles distribute themselves among these states. Boltzmann's distribution predicts the probabilities of particles occupying various energy levels based on temperature and energy differences.

3. Maxwell-Boltzmann Distribution: Velocity Distributions: In gases, the distribution of particle velocities follows the Maxwell-

Boltzmann distribution. It reveals how the speeds of particles vary and how temperature influences this distribution.

4. Thermal Equilibrium: Balancing Act: Boltzmann's distribution applies to systems at thermal equilibrium, where energy exchange between particles is balanced. At equilibrium, particles follow the distribution that maximizes entropy for the given energy constraint.

5. Partition Function: Quantum Insights: The partition function encapsulates the statistical mechanics of a system by summing over all possible states. It provides a bridge between the quantum mechanical properties of particles and macroscopic observables.

6. Applications in Chemistry: Molecular Energies: Boltzmann's distribution underpins concepts in chemical thermodynamics, elucidating how molecules distribute their energy states. It enables the calculation of population ratios for different energy levels.

7. Boltzmann Factor: Thermodynamic Catalysts: The Boltzmann factor quantifies the probability of a system being in a specific energy state. It appears in various contexts, from chemical reactions to molecular motion.

8. Biological Systems: Equilibrium in Living Matter: Boltzmann's distribution applies not only to gases and particles but also to biological systems. It influences protein folding, enzyme reactions, and cellular processes.

9. Applications in Physics and Engineering: Beyond the Molecule: Boltzmann's distribution plays a crucial role in understanding various physical systems, from electrons in a conductor to particles in plasmas. It guides the design of technologies in fields like materials science and electronics.

10. Predicting Behavior: A Glimpse into the Microcosm: Boltzmann's distribution offers a glimpse into the world of particles, revealing how energy and motion intertwine. It equips scientists and engineers with the predictive power to understand and manipulate systems, paving the way for technological innovation and deeper insights into the fundamental nature of matter.

Boltzmann's distribution—this mathematical tapestry that binds energy, probability, and equilibrium—unveils the statistical choreography of particles. It tells us how they waltz among energy levels, embodying the principles that govern both simple gases and complex biological systems. From unraveling the behavior of particles to illuminating the symphony of molecular motion, Boltzmann's distribution guides our exploration of the molecular realm and underpins our quest to comprehend the nature of the universe.

Partition Functions: Quantifying Energy States

In the realm of statistical thermodynamics, partition functions emerge as the architects of molecular behavior, providing a mathematical framework to quantify the distribution of energy states within a system. Rooted in the amalgamation of statistical mechanics and quantum theory, partition functions offer a panoramic view of how energy is apportioned among different states. These functions transcend the microcosmic world of atoms and molecules, guiding our understanding of thermodynamic properties, equilibria, and the intricate dance of particles—a journey that spans from the quantum realm to the macroscopic observables that define the behavior of matter.

1. **Quantum States and Degeneracy: A Multiverse of Possibilities:** In quantum systems, particles can occupy discrete energy states. Some states may have the same energy, leading to degeneracy—the multiplicity of states at a given energy level.

2. **Statistical Mechanics and Probability: Bridging the Gap:** Partition functions bridge the gap between quantum states and macroscopic observables. They encompass the probabilities of particles being in various states, providing the connection to temperature and energy.

3. **Translational, Rotational, and Vibrational States: A Trio of Motion:** In molecules, partition functions account for various types of energy states. Translational states involve motion through space, rotational states involve spinning motion, and

vibrational states involve oscillation of atoms within bonds.

4. Electronic States: Quantum Dance of Electrons: Partition functions also apply to electronic energy levels. They reveal how electrons populate different electronic states, influencing the electronic contributions to thermodynamic properties.

5. Rotational and Vibrational Spectroscopy: A Spectrum of Information: Partition functions find applications in rotational and vibrational spectroscopy. They aid in analyzing spectral data, determining molecular constants, and extracting thermodynamic information.

6. Thermodynamic Properties: Thermodynamic Cartography: Partition functions underpin a range of thermodynamic properties, such as internal energy, entropy, and free energy. These properties offer insights into the system's equilibrium behavior and its response to temperature changes.

7. Equilibrium Constants: Mapping Chemical Reactions: Partition functions facilitate the calculation of equilibrium constants for chemical reactions. They illuminate the balance between reactants and products at different temperatures.

8. Statistical Mechanics and Quantum Chemistry: Harmonizing Theories: Partition functions bridge the gap between statistical mechanics and quantum chemistry. They allow quantum mechanical insights to be connected to thermodynamic observables, providing a comprehensive understanding of molecular behavior.

9. Numerical and Computational Approaches: Calculating Complex Systems: For complex molecules, analytical solutions to partition functions may be challenging. Numerical and computational methods provide powerful tools to estimate partition functions and related properties.

10. Beyond Molecules: Broad Applications: Partition functions extend beyond molecules to gases, solids, and even astrophysical systems. They guide our understanding of diverse systems, from interstellar clouds to the behavior of materials at extreme conditions.

Partition functions—these mathematical architects of energy distributions—paint a portrait of molecular landscapes. They encompass the vibrational symphonies of atoms, the quantum orbits of electrons, and the statistical choreography of particles. As they bridge quantum theory with macroscopic phenomena, partition functions guide our exploration of the energetic intricacies that shape the behavior of matter. From calculating thermodynamic properties to revealing the quantum essence of particles, partition functions unravel the tapestry of energy states, illuminating the fundamental principles that govern the ever-evolving dance of the universe.

Entropy and Equilibrium: The Statistical Underpinnings

In the fascinating realm of thermodynamics, the concepts of entropy and equilibrium emerge as the cornerstones that govern the behavior of matter and energy. Rooted in statistical mechanics and intertwined with the second law of thermodynamics, these principles offer profound insights into the spontaneous direction of processes, the nature of disorder, and the balance achieved in dynamic systems. From the molecular scale to macroscopic observables, the interplay between entropy and equilibrium shapes our understanding of energy transformations, chemical reactions, and the inherent tendency of nature—a journey that spans from microscopic randomness to the macroscopic arrow of time.

1. **Statistical Mechanics and Disorder: The Quest for Macrostates:** Entropy, often dubbed as a measure of disorder, emerges from the microscopic behavior of particles. It quantifies the number of ways particles can be arranged to produce a given macroscopic state.

2. **Boltzmann's Formula: Bridging the Microscopic and Macroscopic:** Boltzmann's entropy formula bridges the gap between the number of microstates and macrostates. It establishes a link between the statistical distribution of particles and the observable properties of a system.

3. **Second Law of Thermodynamics: The Unyielding March of Entropy:** The second law of thermodynamics asserts that the

total entropy of a closed system tends to increase over time. It underpins the notion of irreversibility and the natural direction of processes.

4. Equilibrium and Maximum Entropy: Balancing Act of Nature: At equilibrium, a system's entropy reaches its maximum value given the constraints. This principle of maximum entropy is responsible for the balance achieved in dynamic systems.

5. Chemical Reactions and Gibbs Free Energy: Thermodynamic Compass: Entropy and enthalpy influence chemical reactions. The Gibbs free energy combines these factors and determines whether a reaction is spontaneous (favorable) or non-spontaneous (unfavorable).

6. Entropy and Energy Transfers: Thermodynamic Engines: Entropy considerations guide the design of thermodynamic engines and devices that convert energy from one form to another, such as heat engines and refrigerators.

7. Entropy and Heat: The Dance of Energy: Entropy embodies the concept of heat transfer. Heat naturally flows from higher to lower temperatures, increasing the overall entropy of the system and its surroundings.

8. Entropy in the Universe: The Cosmic Perspective: The universe's entropy tends to increase over time, in accordance with the second law of thermodynamics. This concept ties into cosmological theories about the origin and fate of the universe.

9. Information Theory and Entropy: From Bits to Thermodynamics: Entropy finds applications beyond thermodynamics, including information theory. It quantifies the uncertainty or randomness of data and connects diverse fields.

10. The Arrow of Time: From Past to Future: The increase of entropy aligns with the arrow of time—a concept that distinguishes between the past and the future. The progression of entropy reflects the irreversibility of natural processes.

Entropy and equilibrium—these intertwined concepts—guide our exploration of the behavior of matter and energy. They reveal the underlying principles that dictate the direction of processes, the balance achieved in dynamic systems, and the very essence of change itself. From chemical reactions to the cosmos, entropy and equilibrium illuminate the dynamic symphony of nature, guiding our understanding of the evolution of systems, the passage of time, and the profound interplay between microscopic randomness and the grand tapestry of the universe.

Molecular Simulations: Virtual Experiments in Real Time

In the dynamic realm of computational chemistry, molecular simulations stand as a revolutionary tool that allows us to embark on virtual journeys through the intricate landscapes of molecules and materials. By harnessing the power of computer algorithms, these simulations recreate the behaviors, interactions, and transformations of atoms and molecules in silico, offering insights that complement experimental observations. From elucidating reaction mechanisms to predicting material properties, molecular simulations enable us to explore the molecular world with unprecedented detail and provide a glimpse into the uncharted territories of chemical and physical phenomena—a journey that spans from the principles of molecular dynamics to their transformative applications across scientific disciplines.

1. Molecular Dynamics: Unleashing Atomistic Motion: Molecular dynamics simulations trace the trajectories of individual atoms and molecules as they move and interact over time. They offer a dynamic view of molecular behavior and provide insights into the temporal evolution of systems.

2. Force Fields: Mathematical Representations of Interactions: Force fields encode the interactions between atoms and molecules in simulations. These mathematical models describe the potential energies associated with bond stretching, angle bending, and non-covalent forces.

3. Quantum Mechanical Simulations: Beyond Classical Mechanics: Quantum mechanical simulations delve into the electronic structure of molecules. They offer a more accurate representation of chemical bonding and reactions by solving the Schrödinger equation.

4. Monte Carlo Simulations: Stochastic Exploration: Monte Carlo simulations explore configurational space through random sampling. They are particularly useful for studying thermodynamics, phase transitions, and equilibrium properties.

5. Applications in Drug Discovery: Virtual Screening and Binding: Molecular simulations aid in drug discovery by predicting the binding of small molecules to proteins. They guide the design of potential drug candidates and shed light on the mechanisms of drug-receptor interactions.

6. Materials Science and Nanotechnology: Predicting Properties: Molecular simulations predict material properties and behaviors, facilitating the design of new materials and nanoscale structures with tailored characteristics.

7. Reaction Mechanisms and Catalysis: Unraveling Complex Pathways: Simulations elucidate reaction mechanisms by tracking the sequence of bond-breaking and bond-forming events. They offer a detailed view of how reactions proceed and the roles of catalysts.

8. Protein Folding and Dynamics: Insights into Biomolecules: Molecular simulations explore the intricate process of protein folding and the dynamics of biomolecular systems. They provide insights into the structure-function relationships of proteins.

9. Challenges and Advances: From Complexity to Insight: Molecular simulations face challenges related to accuracy, computational resources, and timescales. Advances in algorithms, hardware, and software are expanding the horizons of simulation capabilities.

10. Interdisciplinary Applications: Crossing Scientific Borders: Molecular simulations transcend disciplinary boundaries, impacting fields from chemistry and physics to biology and materials science. They enable cross-disciplinary collaborations and accelerate scientific discovery.

Molecular simulations—these virtual laboratories—allow us to embark on journeys that were once confined to the imagination. They offer a magnified view of molecular intricacies, revealing the dance of atoms, the transformations of molecules, and the behaviors of materials. From understanding the mechanics of life to designing novel materials, molecular simulations empower scientists to explore the universe of the infinitesimally small, guiding us toward innovations that reshape our world and deepen our comprehension of the molecular fabric that underpins reality.

Monte Carlo and Molecular Dynamics: Sampling Molecular Behavior

In the realm of computational chemistry, Monte Carlo and Molecular Dynamics simulations stand as dynamic windows into the behavior of molecules, offering insights that complement experimental observations. Rooted in statistical mechanics and propelled by the computational prowess of modern technology, these simulations employ algorithms to sample the intricate landscapes of molecular configurations and movements. From exploring thermodynamics to elucidating reaction pathways, Monte Carlo and Molecular Dynamics simulations enable us to delve into the complex choreography of atoms and molecules, unlocking hidden insights and predicting behaviors—a journey that spans from the probabilistic realms of Monte Carlo to the dynamic trajectories of Molecular Dynamics.

1. Monte Carlo Simulations: Stochastic Exploration: Monte Carlo simulations involve random sampling of molecular configurations to explore thermodynamic properties and equilibrium behavior. They excel at calculating averages and probabilities in complex systems.

2. Metropolis Algorithm: The Dance of Acceptance and Rejection: The Metropolis algorithm, a cornerstone of Monte Carlo simulations, guides the acceptance or rejection of proposed molecular configurations. It ensures that configurations are sampled according to the desired probability distribution.

3. Equilibrium Properties: Sampling the Statistical Ensemble:

Monte Carlo simulations yield thermodynamic properties in various statistical ensembles, such as the canonical ensemble (constant temperature), isothermal-isobaric ensemble (constant temperature and pressure), and grand canonical ensemble (constant chemical potential).

4. Applications in Phase Transitions: Solid to Liquid and Beyond: Monte Carlo simulations excel at studying phase transitions, such as the solid-liquid transition. By exploring the behavior of particles at different temperatures and pressures, they unveil the mechanisms behind these transformations.

5. Molecular Dynamics Simulations: Dynamic Atomistic Exploration: Molecular Dynamics simulations track the time evolution of molecules by numerically solving classical equations of motion. They capture dynamic behavior, vibrational motion, and even reaction pathways.

6. Newton's Equations of Motion: Computational Choreography: Newton's equations govern the motion of atoms and molecules in Molecular Dynamics simulations. Forces from interatomic potentials drive the movement, allowing the simulation of molecular trajectories.

7. Time Steps and Integration Algorithms: Taming Molecular Motion: Molecular Dynamics simulations divide time into discrete steps and integrate the equations of motion to predict particle positions. The choice of integration algorithm affects the accuracy and stability of the simulation.

8. Applications in Biomolecular Simulations: Protein Dynamics and Folding: Molecular Dynamics simulations provide insights into biomolecular behavior, such as protein folding, ligand binding, and enzymatic reactions. They bridge the gap between static structural information and dynamic processes.

9. Reaction Pathways and Free Energy Calculations: Unveiling Molecular Transformations: Molecular Dynamics simulations can be used to explore reaction mechanisms and calculate free energy profiles. Advanced techniques like enhanced sampling methods enhance the exploration of rare events.

10. Challenges and Future Directions: From Atoms to Complexity: Both Monte Carlo and Molecular Dynamics simulations face challenges related to system size, timescales, and accuracy. Continued advancements in algorithms, computational power, and techniques like machine learning hold promise for overcoming these challenges.

Monte Carlo and Molecular Dynamics simulations—these computational odysseys—empower us to voyage through the molecular landscape, revealing the vibrational symphonies of atoms, the dynamic dance of molecules, and the transformations that govern chemical and physical phenomena. From predicting material properties to uncovering the secrets of biomolecular behavior, these simulations expand our horizons and guide our exploration of the multifaceted world of matter and motion.

Applications: From Protein Folding to Material Properties

The remarkable capabilities of molecular simulations extend across diverse scientific domains, from unraveling the mysteries of protein folding to predicting the properties of materials. By employing computational tools, scientists can simulate the behaviors of atoms and molecules under various conditions, leading to insights that are often difficult to obtain through experiments alone. These applications span fields such as biology, chemistry, physics, and materials science, transforming our understanding of molecular dynamics, interactions, and reactions—a journey that brings us from the microcosm of biomolecules to the macroscopic world of materials.

1. Protein Folding and Dynamics: Unlocking Biological Mechanisms: Molecular simulations offer unprecedented insights into the intricate process of protein folding. They provide a detailed view of how proteins adopt their three-dimensional structures, aiding in understanding diseases, drug interactions, and functional mechanisms.

2. Drug Design and Binding: Virtual Screening and Ligand Interactions: Simulations aid in drug discovery by predicting how potential drug molecules interact with proteins. They guide the design of new drug candidates and shed light on the intricacies of ligand binding.

3. Enzyme Catalysis: Deciphering Chemical Reactions: Molecular simulations unravel the mechanisms of enzyme-catalyzed reactions. They shed light on the transition states, intermediates, and reaction pathways, aiding in enzyme engineering and rational drug design.

4. Material Properties: Tailoring Novel Materials: Simulations predict the properties of materials by exploring atomic and molecular interactions. They assist in designing materials with specific mechanical, electronic, and thermal properties for applications in electronics, materials science, and nanotechnology.

5. Chemical Reactions and Reaction Mechanisms: Unveiling Transformations: Simulations elucidate reaction mechanisms by tracking the movement of atoms during chemical reactions. They help in understanding reaction pathways, intermediates, and rate-determining steps.

6. Phase Transitions and Thermodynamics: Predicting Transformations: Simulations study phase transitions, such as solid-to-liquid transitions or phase separations. They reveal the thermodynamic behavior of systems at various temperatures and pressures.

7. Protein-Ligand Binding and Drug Discovery: Precision Medicine Insights: Molecular simulations enhance our understanding of protein-ligand interactions, enabling the design of personalized therapies and contributing to the field of precision medicine.

8. Material Stability and Durability: Enhancing Materials Performance: Simulations predict material stability under different conditions, guiding the selection of materials for specific applications and ensuring their durability over time.

9. Nanotechnology and Nanomaterials: Tailoring Nanoscale Structures: Simulations guide the design of nanomaterials with unique properties, such as nanoparticles, nanotubes, and graphene sheets, by exploring their behavior at the nanoscale.

10. Structural Biology and Drug Targeting: Innovations in Treatment: Molecular simulations aid in identifying potential drug targets and designing therapies that specifically interact with these targets, leading to advancements in the treatment of various diseases.

Molecular simulations—these computational laboratories—usher us into realms previously accessible only through imagination. They unravel the dynamic intricacies of biomolecules, predict material properties at the atomic level, and provide a testbed for exploring chemical and physical phenomena. From unveiling the mysteries of protein function to paving the way for novel materials and treatments, molecular simulations empower scientists to forge innovative pathways in fields that span the breadth of scientific inquiry.

Hartree-Fock and Beyond: Solving Schrödinger's Equation

In the intricate landscape of quantum chemistry, the Hartree-Fock method and its extensions stand as pivotal tools for solving the Schrödinger equation—an equation that encapsulates the quantum mechanical behavior of electrons in molecules. Rooted in the principles of quantum mechanics and mathematical approximations, these methods offer a framework to predict molecular properties, energies, and electronic structures. From the early steps of calculating electronic wavefunctions to the sophisticated approaches of correlated wavefunction methods, the Hartree-Fock family of techniques guides us through the enigmatic quantum world, shedding light on the behaviors of atoms and molecules—a journey that spans from single-electron orbitals to the multi-electron interactions that shape chemical and physical phenomena.

1. The Schrödinger Equation: The Quantum Mechanical Blueprint: The Schrödinger equation describes the behavior of electrons in molecular systems. Its solution yields the wavefunctions and energies that characterize the electronic states of molecules.

2. Hartree-Fock Method: Taming Electron-Electron Interactions: The Hartree-Fock method approximates the many-body electron problem by assuming that each electron moves in an average field generated by the other electrons. It provides an efficient way to calculate molecular properties.

3. Self-Consistent Field Iteration: Seeking Consistency: The Hartree-Fock method involves an iterative process that updates the electron density until a self-consistent field is achieved. This iteration improves the accuracy of the calculated wavefunctions.

4. Basis Sets: Mathematical Building Blocks: Basis sets are used to approximate the wavefunctions of electrons. They form the foundation upon which molecular orbitals are constructed and are crucial for accurate calculations.

5. Post-Hartree-Fock Methods: Beyond the Mean Field: Post-Hartree-Fock methods, such as configuration interaction (CI) and coupled cluster (CC) approaches, incorporate electron correlation effects that go beyond the Hartree-Fock approximation.

6. Electron Correlation: The Dance of Electrons: Electron correlation accounts for the interactions between electrons that are not captured in the Hartree-Fock approach. It plays a critical role in accurately describing molecular properties.

7. Density Functional Theory (DFT): The Revolution of Electronic Structure: DFT is a computational approach that models electron density rather than wavefunctions. It offers a balance between accuracy and computational efficiency and is widely used in studying large systems.

8. Hybrid Functionals: Combining Theories: Hybrid functionals combine DFT with Hartree-Fock or other methods, capturing both short-range electron correlation and long-range electron density effects.

9. Excited States and Spectroscopy: Unveiling Electronic Transitions: Extensions of Hartree-Fock methods and DFT enable the calculation of excited states and spectroscopic properties, providing insights into electronic transitions and optical behavior.

10. Challenges and Future Directions: Scaling Quantum Heights:

The Hartree-Fock family of methods faces challenges in treating strongly correlated systems and accurately predicting reaction energies. Advancements in algorithms, parallel computing, and quantum computing hold promise for tackling these challenges.

The Hartree-Fock method and its descendants—these quantum architects—guide us through the labyrinthine quantum world, revealing the symmetries of wavefunctions, the dances of electrons, and the intricacies of molecular properties. From predicting energies to understanding electronic transitions, these methods empower scientists to navigate the microscopic landscapes that govern chemical reactions, material properties, and the very fabric of matter itself.

Density Functional Theory: Predicting Molecular Energies

In the realm of quantum chemistry, Density Functional Theory (DFT) stands as a transformative computational tool, revolutionizing the prediction of molecular energies, properties, and structures. Rooted in the principles of quantum mechanics, DFT offers a pragmatic approach to modeling electronic behavior by focusing on electron density rather than wavefunctions. By harnessing the power of functionals—the mathematical constructs that link electron density to energy—DFT provides a versatile framework to study molecules, materials, and reactions. From simulating chemical reactions to exploring the electronic properties of solids, DFT guides us through the intricate landscape of electronic structure and paves the way for predictive insights into molecular behavior—a journey that bridges theory with computation and observation.

1. Electronic Density: The Crucial Quantity: DFT centers on the electron density—a distribution that characterizes the arrangement of electrons in a system. It encodes information about chemical bonds, interactions, and properties.

2. Hohenberg-Kohn Theorems: Pioneering Principles: The Hohenberg-Kohn theorems form the foundation of DFT. They establish the one-to-one correspondence between electron density and external potential, enabling the formulation of DFT.

3. Kohn-Sham Equations: Simulating Non-Interacting Electrons: The Kohn-Sham equations map the interacting electron system onto an auxiliary system of non-interacting electrons with an effective potential. Solving these equations yields the electron density and energy.

4. Exchange-Correlation Functional: The Missing Link: The exchange-correlation functional encapsulates the elusive interactions of electrons beyond the independent particle model. Approximating this functional is the key challenge in DFT.

5. Local Density Approximation (LDA) and Beyond: LDA is a simple form of the exchange-correlation functional that approximates the energy density as a function of the local electron density. More advanced functionals, like generalized gradient approximations (GGAs), incorporate information about the gradient of the density.

6. Hybrid Functionals: Blending DFT with Hartree-Fock: Hybrid functionals combine DFT with Hartree-Fock methods, offering a balance between computational efficiency and accuracy. They capture both short-range correlation and long-range density effects.

7. Periodic Boundary Conditions: Simulating Solids and Materials: DFT can be extended to study crystalline solids and materials using periodic boundary conditions. It predicts lattice energies, band structures, and material properties.

8. Applications in Reaction Mechanisms: Unveiling Chemical Transformations: DFT predicts reaction energies, activation barriers, and reaction pathways. It aids in understanding reaction mechanisms and catalytic processes.

9. Solvent Effects: Realistic Environments: DFT can incorporate solvent effects by modeling the solvent as a continuum with a dielectric constant. This enables the study of reactions in solution.

10. Challenges and Future Developments: Crossing Quantum Frontiers: DFT has limitations in treating strongly correlated systems and accurately describing long-range dispersion interactions. Advances in functionals, exchange-correlation models, and computational resources hold promise for overcoming these challenges.

Density Functional Theory—this computational paradigm—ushers us into the heart of electronic structure, revealing the fingerprints of atoms, the symphony of electrons, and the energetic signatures that define molecular behavior. From predicting chemical reactivity to simulating material properties, DFT empowers scientists to explore the quantum realm of matter with unprecedented accuracy and predictive power.

Molecular Properties: Insights from Quantum Calculations

In the realm of quantum chemistry, the power of quantum calculations extends beyond predicting molecular energies—it delves into a rich tapestry of molecular properties that offer deep insights into the behavior, interactions, and reactivity of atoms and molecules. Leveraging the principles of quantum mechanics and computational algorithms, these calculations provide a window into the molecular world, enabling scientists to unravel the intricacies of chemical bonding, spectroscopic behavior, and electronic structure. From understanding the geometric arrangement of atoms to deciphering the origins of molecular spectra, quantum calculations guide us through the diverse landscape of molecular properties—a journey that bridges the microcosm of electrons to the macroscopic world of experimental observations.

1. Geometric Parameters: Probing Molecular Structures: Quantum calculations determine molecular geometries—bond lengths, bond angles, and dihedral angles—that shape molecular properties and reactivity. They provide a theoretical foundation for structural analysis.

2. Dipole Moments: Polar Insights into Molecular Charge: Dipole moments reveal the charge distribution within molecules. Quantum calculations predict dipole moments, helping understand molecular polarity and interactions.

3. Vibrational Frequencies: Simulating Molecular Vibrations:

Quantum calculations predict vibrational frequencies, aiding in understanding molecular vibrations, infrared spectra, and the characterization of functional groups.

4. Electronic Spectra: Unraveling Molecular Absorption: Quantum calculations predict electronic spectra, shedding light on the absorption of light by molecules and the origins of UV-Vis and fluorescence spectra.

5. NMR Shielding and Chemical Shifts: Navigating Magnetic Resonance: Quantum calculations predict nuclear magnetic resonance (NMR) shielding constants, aiding in the interpretation of NMR spectra and providing insights into chemical environments.

6. Electron Density and Electrostatic Potential: Mapping Charge Distribution: Quantum calculations yield electron density and electrostatic potential maps, visualizing the charge distribution within molecules and aiding in understanding intermolecular interactions.

7. Molecular Orbitals: Bridging Structure and Reactivity: Quantum calculations predict molecular orbitals, elucidating the electronic structure of molecules and guiding our understanding of chemical reactivity.

8. Transition States and Reaction Mechanisms: Unveiling Activation Barriers: Quantum calculations identify transition states and activation energies, illuminating reaction mechanisms and pathways, and providing insights into reaction kinetics.

9. Solvent Effects: Modeling Molecular Environments: Quantum calculations with solvent models simulate the influence of solvents on molecular properties, making them applicable to reactions in solution.

10. Computational Accuracy and Challenges: Navigating Quantum Realms: Quantum calculations face challenges related to computational resources, approximations, and the treatment of electron correlation. Advances in methodologies and techniques continue to enhance accuracy.

Molecular properties—these quantum fingerprints—guide us through the intricate realm of matter, offering insights that resonate from the atomic to the macroscopic scales. Quantum calculations empower scientists to unravel the mysteries of molecular structure, reactivity, and behavior, bridging the gap between theory and experiment, and fostering a deeper understanding of the molecular universe that surrounds us.

Redox Reactions: Electron Transfer Events

Redox reactions, characterized by the exchange of electrons between chemical species, lie at the heart of numerous chemical processes, from energy generation to biological functions. These electron transfer events drive the transformations of matter and underpin the principles of oxidation and reduction. Redox reactions encompass a wide spectrum of phenomena, from simple electron transfers to complex enzymatic processes. Through the lens of electron exchange, we can explore the interplay of chemical species, the transfer of energy, and the balance of oxidation states—a journey that spans from the reactivity of metals to the intricacies of biological redox signaling.

1. Oxidation and Reduction: Balancing Electron Exchanges: Oxidation involves the loss of electrons, leading to an increase in oxidation state, while reduction entails the gain of electrons, resulting in a decrease in oxidation state.

2. Half-Reactions: Bridging Electron Donors and Acceptors: Redox reactions can be divided into half-reactions, each describing the electron donor and acceptor. The balancing of these half-reactions ensures the conservation of electrons.

3. Electrochemical Cells: Harnessing Electron Flow: Electrochemical cells use redox reactions to convert chemical energy into electrical energy and vice versa. Batteries and fuel cells exemplify the utilization of redox chemistry for practical applications.

4. Standard Electrode Potentials: Quantifying Redox Tendencies: Standard electrode potentials measure the tendency of a species to donate or accept electrons. They provide insight into the spontaneity of redox reactions.

5. Nernst Equation: Predicting Cell Potentials: The Nernst equation relates the electrode potential of a redox couple to the concentrations of reactants and products, enabling the prediction of cell potentials under non-standard conditions.

6. Balancing Redox Reactions: Methodologies and Equations: Redox reactions are balanced using techniques like the oxidation number method and the half-reaction method. These methods ensure the conservation of mass and charge.

7. Biological Redox Reactions: Life's Energy Currency: Biological systems rely on redox reactions for processes like cellular respiration and photosynthesis. The transfer of electrons fuels the synthesis of ATP, the energy currency of life.

8. Redox Signaling: Balancing Cellular Equilibrium: Redox signaling involves the regulation of cellular processes through the controlled release of reactive oxygen species (ROS) and redox-sensitive molecules.

9. Electron Transfer in Metals: Corrosion and Electroplating: Redox reactions play a role in metal corrosion and electroplating. Corrosion involves the oxidation of metals, while electroplating involves the reduction of metal ions onto a surface.

10. Environmental Implications: Redox in Natural Systems: Redox reactions influence soil chemistry, groundwater contamination, and nutrient cycling in natural environments, impacting ecosystems and human activities.

Redox reactions—these electron choreographies—shape our understanding of energy transformations, chemical reactivity, and the intricate balance of electron exchanges. From the powerhouse of cellular respiration to the rusting of metals, these reactions drive the dynamics of matter and energy across scales, unveiling the vital role they play in chemistry, biology, and our interactions with the natural world.

Electrochemical Cells: Harnessing Electron Flow

Electrochemical cells stand as versatile devices that ingeniously exploit redox reactions to transform chemical energy into electrical energy and vice versa. These cells bridge the realms of chemistry and electricity, enabling a diverse array of applications, from powering electronic devices to facilitating industrial processes. By coupling oxidation and reduction half-reactions, electrochemical cells allow for controlled electron flow and provide a conduit for the conversion of chemical potential energy into a usable form. Through the interplay of electrodes, electrolytes, and redox species, electrochemical cells revolutionize our ability to generate power, store energy, and drive chemical transformations—a journey that encompasses galvanic cells, electrolytic cells, and beyond.

1. Galvanic Cells: Spontaneous Electron Flow: Galvanic cells, also known as voltaic cells, spontaneously generate electrical energy from a chemical reaction. They consist of two half-cells connected by a conductive pathway, allowing electrons to flow from the anode to the cathode.

2. Cell Potential: Measuring Energy Conversion: The cell potential, or electromotive force (EMF), quantifies the ability of an electrochemical cell to drive electron flow. It is related to the standard electrode potentials of the half-reactions involved.

3. Salt Bridge: Maintaining Charge Neutrality: A salt bridge connects the two half-cells in a galvanic cell, allowing ions to migrate and maintain charge neutrality as electrons flow. This prevents an electrostatic buildup that would impede the cell's operation.

4. Electrolytic Cells: Forcing Non-Spontaneous Reactions: Electrolytic cells, distinct from galvanic cells, require an external source of energy to drive non-spontaneous reactions, such as electrolysis or electroplating. They allow the control of chemical transformations using electrical energy.

5. Faraday's Laws of Electrolysis: Quantifying Electrochemical Reactions: Faraday's first and second laws establish the relationship between the amount of substance transformed in an electrolytic cell and the quantity of charge passed through the cell.

6. Fuel Cells: Clean Energy Conversion: Fuel cells operate like galvanic cells but continuously replenish reactants to generate electricity. Hydrogen fuel cells, for instance, combine hydrogen and oxygen to produce water and electrical energy, with only water and heat as byproducts.

7. Battery Technologies: Portable Energy Storage: Batteries are composed of interconnected electrochemical cells. They store electrical energy for later use and play a pivotal role in portable electronic devices, electric vehicles, and renewable energy systems.

8. Redox Flow Batteries: Scalable Energy Storage: Redox flow batteries store energy in liquid electrolytes housed in external tanks. They are scalable and well-suited for grid energy storage applications.

9. Applications in Industry: Electrosynthesis and Corrosion Protection: Electrochemical cells are used for electrosynthesis, where chemical reactions are driven by electricity. They also find applications in corrosion protection and metal electroplating.

10. Environmental Impacts and Sustainability: Green Energy Solutions: Electrochemical cells contribute to sustainable energy solutions by enabling renewable energy integration and reducing reliance on fossil fuels.

Electrochemical cells—these dynamic junctions of chemical and electrical realms—power our technological advancements and pave the way toward a cleaner, more sustainable future. From lighting up our lives to driving the engines of progress, they transform the landscape of energy conversion and storage, demonstrating the ingenious synergy between chemical reactions and electron flow.

Nernst Equation: Electrochemical Equilibrium

The Nernst equation stands as a fundamental tool in electrochemistry, illuminating the relationship between the electrode potential of an electrochemical cell and the concentrations of reactants and products involved in the redox reactions. Named after the German physicist Walther Nernst, this equation offers insights into the driving force of electron transfer, the spontaneity of redox reactions, and the conditions required for achieving electrochemical equilibrium. By quantifying the impact of concentration changes on cell potential, the Nernst equation serves as a guide for predicting and understanding the behavior of electrochemical systems—a journey that spans from standard electrode potentials to the non-ideal world of real electrochemical cells.

1. Standard Electrode Potential: Reference for Electrochemical Reactions: The standard electrode potential ($E°$) represents the potential of an electrode in a half-cell compared to a standard hydrogen electrode (SHE). It is a measure of the tendency of a redox couple to gain or lose electrons.

2. Nernst Equation Derivation: Quantifying Concentration Effects: The Nernst equation arises from the application of thermodynamic principles to electrochemical systems. It relates the cell potential (E) to the standard electrode potential, the gas constant (R), temperature (T), the number of electrons transferred (n), and the concentrations of reactants and products.

3. Concentration Cell: Balancing Concentrations and Potentials: A concentration cell consists of two half-cells with the same electrode materials but different concentrations of the same redox couple. The Nernst equation predicts that the cell potential becomes zero at equilibrium.

4. Non-Ideal Behavior: Considering Activity and Real Electrodes: In real electrochemical systems, the Nernst equation accounts for non-ideal behavior through the concept of activity, which corrects for deviations from ideal behavior due to ion interactions.

5. Nernst Equation and pH: pH-Dependent Electrode Potentials: The Nernst equation can be applied to pH electrodes, such as the glass electrode, which measures pH based on the potential difference between a reference electrode and a pH-sensitive electrode.

6. Biological Applications: Ion Concentration and Cellular Function: The Nernst equation finds applications in understanding ion transport across cell membranes, nerve cell action potentials, and pH regulation in biological systems.

7. Applications in Analytical Chemistry: Potentiometric Measurements: The Nernst equation is used in potentiometric measurements to determine ion concentrations in solutions, such as pH measurements using glass electrodes.

8. Limitations and Extensions: Beyond the Ideal Scenario: The Nernst equation assumes ideal conditions and does not consider factors like electrode impedance and activity coefficients. Extended versions of the equation incorporate these factors for more accurate predictions.

9. Predicting Equilibrium: Electrochemical Equilibrium Conditions: The Nernst equation helps predict the conditions under which an electrochemical cell reaches equilibrium, offering insights into the behavior of redox reactions.

10. Industrial and Environmental Implications: Monitoring and Control: The Nernst equation is employed in various industries for monitoring and controlling electrochemical processes, such as corrosion prevention and wastewater treatment.

The Nernst equation—this mathematical bridge between concentration gradients and electrode potentials—unveils the delicate interplay of chemical species in electrochemical systems. From predicting equilibrium conditions to powering analytical instruments, this equation underpins our ability to harness and understand electron transfer events, enabling advancements in fields ranging from biology to industry.

Ligand-Field Theory: Understanding Metal Complexes

Ligand-Field Theory stands as a cornerstone in the realm of inorganic chemistry, providing a comprehensive framework for understanding the electronic structure and properties of transition metal complexes. By focusing on the interactions between transition metal ions and surrounding ligands, this theory illuminates the intricate coordination chemistry that shapes the behavior of metal complexes. Through the lens of ligand-field theory, we can unravel the origin of color, magnetic properties, and reactivity of coordination compounds—a journey that navigates the crystal field, splitting of d orbitals, and the multifaceted world of spectroscopy and ligand effects.

1. Transition Metal Complexes: Coordination Chemistry Unveiled: Transition metal complexes comprise a central metal ion surrounded by coordinating ligands. Ligand-Field Theory explains their electronic structure and properties.

2. Crystal Field Theory: Electrostatic Interactions and Splitting: Crystal Field Theory approximates ligands as point charges, leading to the splitting of d orbitals into different energy levels, known as crystal field splitting.

3. Ligand Coordination and Spectrochemical Series: Tuning Crystal Field Effects: Different ligands exert varying degrees of crystal field splitting, influencing the color and magnetic properties of metal complexes. The spectrochemical series ranks ligands based on their strength.

4. Octahedral and Tetrahedral Complexes: Shape and Symmetry: In octahedral complexes, ligands surround the metal in a symmetrical arrangement. Tetrahedral complexes exhibit a different arrangement, leading to distinct crystal field splittings.

5. Crystal Field Stabilization Energy (CFSE): Explaining Stability: CFSE quantifies the energy difference between the lower and higher energy sets of d orbitals due to ligand-field interactions. It explains the stability of certain coordination numbers and geometries.

6. Ligand-Field Theory and Color: The Origin of Complex Colors: The absorption of light by metal complexes arises from electronic transitions between d orbitals split by crystal field interactions. Ligand-Field Theory explains the origins of complex colors.

7. Spin Crossover and Magnetic Properties: The Role of Ligand-Field Effects: Ligand-Field Theory elucidates the magnetic properties of metal complexes, including spin crossover phenomena and the coupling of unpaired electrons.

8. Jahn-Teller Effect: Distortions and Electronic Degeneracy: The Jahn-Teller effect explains distortions in some metal complexes due to the unequal occupation of degenerate d orbitals, leading to symmetry-breaking geometries.

9. Spectroscopy of Metal Complexes: Probing Ligand-Field Effects: Various spectroscopic techniques, such as UV-Vis, IR, and EPR spectroscopy, provide insights into ligand-field effects and electronic transitions in metal complexes.

10. Ligand Effects and Reactivity: Modulating Reactivity with Ligand Choice: Different ligands influence the reactivity of metal complexes. Ligand-Field Theory helps predict how ligands impact the stability, reactivity, and catalytic properties of complexes.

Ligand-Field Theory—this captivating exploration of metal-ligand interactions—offers a unified framework for deciphering the electronic intricacies of transition metal complexes. From understanding the vibrant colors of coordination compounds to predicting their reactivity, this theory illuminates the chemical symphony that orchestrates the behavior of metal complexes, transcending disciplines and revealing the elegant interplay of coordination chemistry and spectroscopy.

Electron Transfer and Coordination Chemistry

Electron transfer processes lie at the heart of coordination chemistry, sculpting the reactivity and properties of transition metal complexes. The interplay between electron transfer events and the coordination sphere gives rise to a rich tapestry of redox reactions, ligand effects, and catalytic processes. From metalloenzymes to molecular electronics, electron transfer in coordination compounds underpins a diverse array of applications and phenomena. Exploring this landscape reveals the dynamic dance of electrons, ligands, and metal centers—a journey that spans from redox-active ligands to photoinduced electron transfer and beyond.

1. Ligand-Mediated Electron Transfer: Redox-Active Ligands: Ligands can influence electron transfer by participating in redox reactions themselves. Redox-active ligands play a vital role in electron storage and transport.

2. Ligand-to-Metal Electron Transfer (LMET): Controlling Redox States: In LMET, ligands transfer electrons to or from metal centers, altering the metal's oxidation state. This can lead to new reactivity and electronic configurations.

3. Photoinduced Electron Transfer: Harnessing Light Energy: Photoinduced electron transfer involves the transfer of electrons upon absorption of light. It is crucial in photosynthesis, photovoltaic devices, and artificial photosynthesis.

4. Electron Transfer in Metalloenzymes: Nature's Catalysts:

Metalloenzymes, such as cytochromes and hemoproteins, utilize electron transfer to carry out vital biological functions like oxygen transport and electron transport chains.

5. Mixed-Valence Compounds: Bridging Electron Delocalization: Mixed-valence compounds exhibit different oxidation states on multiple metal centers connected by bridging ligands. They showcase electron delocalization and communication.

6. Inner-Sphere and Outer-Sphere Electron Transfer: Pathways and Mechanisms: Inner-sphere electron transfer involves direct coordination of reactants, while outer-sphere transfer occurs through solvent or ligand-mediated pathways. These mechanisms impact reaction rates.

7. Marcus Theory: Quantifying Electron Transfer Rates: Marcus theory provides a quantitative framework to understand electron transfer rates by considering the energy difference between reactants and products and the reorganization energy.

8. Ligand Effects on Electron Transfer: Tuning Reactivity with Ligand Design: Different ligands influence the energetics and pathways of electron transfer reactions, enabling fine-tuning of reactivity and selectivity.

9. Electron Transfer in Catalysis: Redox Mediators and Pathways: Electron transfer processes are central to catalytic cycles in many metal complexes. Redox mediators facilitate electron transfer between the catalyst and substrate.

10. Molecular Electronics: Coordination Compounds as Conductors: Coordination compounds with metal centers can act as molecular wires in molecular electronics, enabling charge transport in nanoscale devices.

Electron transfer in coordination chemistry—this mesmerizing ballet of charges—shapes the behavior of metal complexes and extends its influence to fields ranging from biochemistry to materials science. From mimicking nature's redox enzymes to constructing molecular devices, this dance of electrons transcends disciplines, revealing the intricate symphony of coordination chemistry's interplay with electron transfer phenomena.

Crystal Field Theory: Geometry and Spectroscopy

Crystal Field Theory (CFT) stands as a foundational concept in inorganic chemistry, providing a powerful framework for understanding the electronic structure, geometry, and spectroscopic properties of transition metal complexes. By considering the electrostatic interactions between metal ions and surrounding ligands, CFT delves into the origins of color, magnetic behavior, and reactivity exhibited by coordination compounds. From octahedral to tetrahedral complexes, this theory unveils the geometric consequences of ligand binding and offers insights into the intricate dance of electrons—a journey that navigates crystal field splitting, spectroscopy, and the elegant interplay of electronic transitions.

1. Geometry and Ligand Arrangements: Defining Complex Shapes: CFT elucidates how ligands influence the arrangement of d orbitals and geometries of transition metal complexes, including octahedral and tetrahedral coordination.

2. Crystal Field Splitting: Energy Levels and Degeneracy: The electrostatic interaction between ligands and metal ions leads to the splitting of degenerate d orbitals into different energy levels—a phenomenon known as crystal field splitting.

3. Spectrochemical Series: Ranking Ligand Strength: CFT predicts how ligands influence crystal field splitting and organizes ligands in the spectrochemical series based on their ability to cause different splitting patterns.

4. Color and Electronic Transitions: The Color of Coordination Compounds: CFT explains how the absorption of light corresponds to electronic transitions between split d orbitals, offering insights into the colors displayed by metal complexes.

5. Magnetic Properties: Spin and Magnetism: CFT reveals how the number of unpaired electrons in a metal complex's d orbitals influences its magnetic properties and the potential for paramagnetism or diamagnetism.

6. High-Spin and Low-Spin Configurations: Balancing Energy and Spin: CFT predicts whether a complex adopts a high-spin or low-spin electronic configuration based on the relative energies of its d orbitals and ligand field splitting.

7. Spectroscopy of Transition Metal Complexes: Probing Electronic Transitions: CFT helps interpret various spectroscopic techniques, such as UV-Vis, IR, and EPR spectroscopy, by relating electronic transitions to crystal field splitting.

8. Limitations and Extensions: Beyond Electrostatic Interactions: CFT has limitations, such as not considering covalent bonding effects and neglecting the electron-electron repulsion interactions in d orbitals. Ligand Field Theory (LFT) extends CFT to address these aspects.

9. Spectroscopic Consequences of Ligand Types: Ligand-Field Effects: Different ligands produce distinct crystal field splitting patterns, which in turn influence the electronic and spectroscopic properties of metal complexes.

10. Insights into Reactivity and Bonding: Ligand Field and Chemical Behavior: CFT provides insights into the reactivity of transition metal complexes, shedding light on their bonding patterns, stability, and potential for catalysis.

Crystal Field Theory—this elegant dance of electrons within the confines of ligand interactions—unveils the geometric landscapes and spectral fingerprints of transition metal complexes. From deciphering complex colors to predicting magnetic behaviors, this theory continues to shape our understanding of the electronic symphony orchestrated by coordination compounds, revealing the profound connection between geometry, spectroscopy, and chemical reactivity.

Reaction Mechanisms: Unraveling Organic Transformations

Reaction mechanisms provide a window into the world of organic chemistry, unveiling the intricate pathways through which molecules transform and bonds break and form. These step-by-step narratives illuminate the roles of reactive intermediates, transition states, and the forces that guide chemical reactions. By deciphering reaction mechanisms, chemists gain insights into the factors influencing reactivity, selectivity, and the stereochemistry of organic transformations—a journey that spans from nucleophilic substitutions to radical reactions and beyond.

1. Reaction Mechanism Basics: Beyond Chemical Equations: Reaction mechanisms go beyond balanced chemical equations, detailing the sequence of events and the intermediates involved in a chemical transformation.

2. Reaction Intermediates: Reactive Stepping Stones: Intermediates are short-lived species formed and consumed during a reaction. They often provide a glimpse into the reaction's progress.

3. Transition States: Crossing Energy Barriers: Transition states represent fleeting arrangements of atoms at energy maxima along reaction pathways. They dictate reaction rates and energies.

4. Bond Breaking and Formation: Unraveling Molecular Transformations: Reaction mechanisms shed light on how bonds are broken and formed during a reaction, influencing the overall outcome and products.

5. Nucleophilic Substitution: SN1 and SN2 Pathways: Nucleophilic substitution reactions involve the replacement of one nucleophile by another. SN1 and SN2 pathways offer distinct mechanisms with different reaction kinetics and stereochemistry.

6. Elimination Reactions: Removing and Rearranging Atoms: Elimination reactions lead to the removal of atoms or groups from a molecule, often generating double bonds. E1 and E2 pathways showcase different mechanisms and stereochemistry.

7. Addition Reactions: Merging Molecules: Addition reactions involve the addition of two or more molecules to form a single product. They play a pivotal role in organic synthesis and the creation of complex molecules.

8. Radical Reactions: Chain Propagation and Termination: Radical reactions involve the formation and reactivity of radical intermediates. Chain propagation, termination steps, and radical stability influence the outcome.

9. Rearrangement Reactions: Atom Shuffle: Rearrangement reactions involve the migration of atoms or groups within a molecule, leading to structural changes. They often involve carbocation intermediates.

10. Stereochemistry and Reaction Mechanisms: Asymmetry Matters: Reaction mechanisms provide insights into the stereochemistry of products, elucidating how reactants' spatial arrangements affect outcomes.

11. Reaction Mechanism Determination: Experimental and Computational Approaches: Experimentation and computational methods, such as quantum mechanics, play essential roles in elucidating reaction mechanisms.

12. Catalysis and Reaction Pathways: Accelerating Transformations: Catalysts influence reaction mechanisms by providing alternative pathways with lower energy barriers, increasing reaction rates, and controlling selectivity.

Unraveling reaction mechanisms—this intricate journey into molecular choreography—guides chemists through the dynamic world of organic transformations. From understanding bond breaking to predicting stereochemical outcomes, these narratives illuminate the underlying forces that shape chemical reactivity and enable the precision of modern synthetic chemistry.

Transition States and Reaction Kinetics

Transition states, the fleeting moments of highest energy along reaction pathways, hold the key to understanding reaction kinetics—the study of reaction rates and mechanisms. By delving into the energetics of transition states, chemists can predict the speed at which reactions occur, determine rate laws, and unravel the intricacies of chemical transformations. This journey through reaction kinetics navigates activation energy, rate-determining steps, and the dynamic interplay of reactants and products, shedding light on the temporal aspects of chemical reactions.

1. Transition State Theory: Energetic Peaks of Reactivity: Transition states represent energy maxima along reaction coordinates. Transition State Theory connects these states to reaction rates, offering insights into activation barriers.

2. Activation Energy: Overcoming Energy Barriers: Activation energy is the energy difference between reactants and the transition state. It determines the pace of reactions and their temperature dependence.

3. Reaction Rates and Rate Laws: Expressing Reaction Speed: Rate laws describe the relationship between reactant concentrations and reaction rates. They reveal how reaction rates change with reactant concentrations.

4. Rate-Determining Steps: Bottlenecks and Mechanistic Insights: The slowest step in a reaction mechanism, known as the rate-determining step, controls the overall rate of the reaction and shapes its kinetics.

5. Reaction Mechanisms and Elementary Steps: Breaking Down

Complex Reactions: Complex reactions can often be dissected into a sequence of elementary steps. The overall rate depends on the slowest elementary step—the rate-determining step.

6. Reaction Order and Molecularity: Examining Rate Laws: Reaction order indicates how reactant concentrations affect reaction rates. Molecularity specifies the number of reactant molecules involved in an elementary step.

7. Temperature and Reaction Rates: The Kinetics of Heat: The Arrhenius equation relates temperature and activation energy to reaction rates, showcasing the role of temperature in controlling reaction kinetics.

8. Catalysis and Reaction Rates: Catalysts as Reaction Accelerators: Catalysts alter reaction pathways, lowering activation barriers and accelerating reactions. They enable alternative routes that enhance reaction rates.

9. Reaction Kinetics in Industry: Efficient Processes and Design: Reaction kinetics underpins the design of efficient industrial processes, allowing engineers to optimize reaction conditions and maximize product yield.

10. Reaction Kinetics in Biological Systems: Enzymes and Cellular Dynamics: Enzymes in biological systems catalyze reactions with remarkable specificity and efficiency, highlighting the importance of reaction kinetics in cellular processes.

11. Transition State Analogs: Designing Enzyme Inhibitors: Transition state analogs mimic the transition state of a reaction and can serve as potent inhibitors of enzymes, offering insights into drug design.

12. Dynamic Kinetics: Beyond Equilibrium: Dynamic kinetics explore reactions that don't achieve equilibrium and delve into the complex interplay of multiple reaction pathways.

13. Kinetic Isotope Effects: Probing Reaction Mechanisms:

Isotope effects in reaction rates provide insights into reaction mechanisms by revealing the role of specific atoms in the transition state.

Exploring transition states and reaction kinetics—this kinetic voyage through the energy landscapes of chemical transformations—illuminates the temporal aspects of chemical reactions. From activation energy to rate-determining steps, this journey not only predicts reaction rates but also unveils the intricate choreography of molecules as they navigate the complex terrain of chemical change.

Solvent Effects and Stereochemistry

Solvents are more than just passive reaction media; they can significantly influence reaction outcomes, including stereochemistry—the spatial arrangement of atoms in molecules. Understanding solvent effects on stereochemistry is essential for controlling the formation of chiral molecules and achieving specific product distributions. This exploration delves into the role of solvents in influencing reaction rates, equilibrium constants, and the formation of stereoisomers—a journey that uncovers the dynamic interplay of molecules, solvents, and chirality.

1. Solvent-Solute Interactions: The Solvent as a Reaction Partner: Solvents can interact with reactants and products through hydrogen bonding, dipole-dipole interactions, and van der Waals forces, affecting reaction pathways.

2. Solvent Polarity and Ionization: Dissolving Ions and Polar Molecules: Polar solvents can stabilize ions and polar transition states, influencing the rates and mechanisms of reactions.

3. Hydrogen-Bonding Solvents: Influence on Reaction Mechanisms: Hydrogen-bonding solvents can affect reaction mechanisms by stabilizing or destabilizing specific intermediates or transition states.

4. Solvent Effects on Reaction Rates: Facilitating or Slowing Reactions: Solvent polarity and interactions can impact reaction rates by stabilizing or destabilizing transition states, leading to changes in activation energy.

5. Stereoselective Solvation: Chiral Environments and

Stereocenters: Chiral solvents or achiral solvents with chiral additives can influence the formation of specific stereoisomers, enabling stereoselective synthesis.

6. Solvent Effects on Equilibrium Constants: Shifting Reaction Equilibria: Solvent interactions with reactants and products can affect reaction equilibria, leading to shifts in product distributions.

7. Solvent Effects on Transition States: Distorting Activation Barriers: Solvent molecules surrounding a transition state can distort its geometry, affecting activation energy and reaction rates.

8. Chirality Transfer: Solvent-Induced Stereoselectivity: Certain solvents can induce chirality transfer, affecting the enantiomeric excess and favoring the formation of specific enantiomers.

9. Supramolecular Solvent Effects: Templating Chiral Structures: Supramolecular interactions between solvents and reactants can influence the formation of chiral structures and control stereochemistry.

10. Solvent Effects in Enzymatic Reactions: Nature's Catalysts and Solvents: Enzymes, as nature's catalysts, often create a specific solvent environment that contributes to their high stereoselectivity and efficiency.

11. Solvent-Induced Conformational Changes: Dynamic Solvent-Solute Interplay: Solvent interactions can induce conformational changes in reactants or products, leading to shifts in their stereochemistry.

12. Solvent Effects in Asymmetric Synthesis: Controlling Chirality: Asymmetric synthesis relies on precise control of stereochemistry, and solvent selection can significantly impact the enantiomeric excess of products.

Solvent effects on stereochemistry—this dynamic interplay

between solvents and molecular spatial arrangements—reveals the profound impact that the surrounding environment can have on chemical reactions. From chirality transfer to reaction rates, this journey showcases the versatility of solvents as active participants in the chemical symphony, enabling chemists to fine-tune reactions and achieve desired stereochemical outcomes.

Nanotechnology: Manipulating Matter at the Nanoscale

Nanotechnology stands as a groundbreaking frontier that empowers scientists and engineers to manipulate matter at the nanoscale—where materials exhibit unique properties and behaviors. This multidisciplinary field harnesses the principles of physics, chemistry, biology, and engineering to design and create structures, devices, and materials with unprecedented control and precision. From nanomaterials to nanomedicine, this journey delves into the diverse facets of nanotechnology, unveiling the remarkable potential to transform industries, revolutionize healthcare, and reshape our understanding of the material world.

1. Introduction to Nanotechnology: Miniaturization and Innovation: Nanotechnology involves the manipulation and control of materials and structures at the nanoscale, often leading to new and enhanced properties.

2. Nanomaterials: Properties Beyond the Macroscopic: Materials at the nanoscale exhibit unique properties, such as enhanced surface area, quantum effects, and altered electronic and mechanical behavior.

3. Bottom-Up and Top-Down Approaches: Crafting Nanostructures: Bottom-up approaches assemble nanoscale structures from individual molecules, while top-down approaches carve larger structures down to the nanoscale.

4. Nanoparticles: Building Blocks of Nanotechnology: Nanoparticles serve as versatile building blocks with applications ranging from drug delivery to catalysis, utilizing their size-dependent properties.

5. Nanodevices: Functionalizing at the Nanoscale: Nanodevices integrate nanomaterials into functional systems, enabling applications in electronics, sensing, and energy storage.

6. Nanomedicine: Transforming Healthcare: Nanotechnology has revolutionized medicine by enabling targeted drug delivery, imaging, and diagnostics at the molecular level.

7. Nanoelectronics: Shrinking Circuits, Expanding Possibilities: Nanoelectronics push the limits of miniaturization in electronic devices, leading to faster, more efficient, and energy-saving technologies.

8. Nanotechnology in Energy: Harnessing Power at the Nanoscale: Nanotechnology plays a pivotal role in energy conversion, storage, and harvesting through innovations in solar cells, batteries, and fuel cells.

9. Nanoscale Characterization: Seeing the Unseen: Advanced techniques like scanning probe microscopy and electron microscopy enable the visualization and analysis of nanoscale structures.

10. Environmental Nanotechnology: Addressing Challenges and Sustainability: Nanotechnology offers solutions for environmental challenges, including pollution remediation, water purification, and efficient energy usage.

11. Ethical and Societal Implications: Navigating the Nanoworld: Nanotechnology raises ethical, regulatory, and safety concerns that must be addressed to ensure responsible development and application.

12. Nanotechnology's Future: Impact and Innovation: Nanotechnology's potential spans diverse fields, from computing to medicine to materials science, promising continued innovation and transformation.

Nanotechnology—this realm of discovery and innovation—unveils the potential to engineer materials and systems at a scale where the properties of matter diverge from the macroscopic world. From redefining healthcare to advancing energy technologies, this journey showcases the transformative power of manipulating matter at the nanoscale, bridging scientific disciplines and reshaping the boundaries of what is possible.

Quantum Computing: Reimagining Computational Power

Quantum computing represents a paradigm shift in the world of computation, harnessing the unique properties of quantum mechanics to process information in ways that classical computers cannot. This revolutionary technology promises exponential speedups for complex calculations, disrupting fields such as cryptography, optimization, and material science. This exploration delves into the principles of quantum mechanics, quantum bits (qubits), quantum gates, and quantum algorithms, revealing how quantum computing redefines the limits of computational power and opens new frontiers of scientific and technological advancement.

1. Introduction to Quantum Computing: Beyond Classical Limits: Quantum computing leverages the principles of quantum mechanics to process information in novel ways, potentially solving problems that are intractable for classical computers.

2. Quantum Bits (Qubits): The Quantum Currency: Qubits are the fundamental units of quantum information, representing the quantum analog of classical bits. Their superposition and entanglement create the foundation for quantum computing.

3. Quantum Gates: Manipulating Qubits: Quantum gates perform operations on qubits, enabling complex computations through quantum parallelism and entanglement.

4. Quantum Entanglement: Nonclassical Correlations: Entanglement links qubits in nonclassical correlations, enabling instantaneous communication and enhancing computational power.

5. Quantum Algorithms: Solving Problems Faster: Quantum algorithms, such as Shor's algorithm for factoring large numbers and Grover's algorithm for searching unsorted databases, promise exponential speedups for specific problems.

6. Quantum Error Correction: Overcoming Decoherence: Quantum computers are sensitive to noise and errors due to interactions with the environment. Quantum error correction techniques mitigate these challenges.

7. Applications of Quantum Computing: Cryptography, Optimization, and More: Quantum computing has the potential to revolutionize cryptography by breaking classical encryption methods and accelerate optimization problems in fields like logistics and finance.

8. Quantum Simulations: Exploring Complex Systems: Quantum computers can simulate quantum systems, providing insights into molecular interactions, materials properties, and chemical reactions.

9. Quantum Supremacy: Crossing the Computational Frontier: Quantum supremacy refers to the point at which a quantum computer outperforms classical computers in specific tasks, demonstrating the practical advantage of quantum computing.

10. Challenges and Future Directions: Building Quantum Computers: Building and maintaining stable quantum systems, known as quantum coherence, remains a significant challenge. Quantum computing research continues to explore new hardware and algorithms.

11. Quantum Ethics and Impact: Navigating Societal Implications: As quantum computing advances, it raises ethical, security, and privacy concerns that must be addressed to ensure responsible and equitable development.

12. Quantum Computing's Promise: Transforming Science and Technology: Quantum computing holds the potential to revolutionize industries, from cryptography and finance to drug discovery and artificial intelligence, ushering in a new era of computational power.

Quantum computing—this marriage of quantum mechanics and computation—ushers in an era of computational power that transcends classical limitations. From reshaping cryptography to accelerating scientific discovery, this journey illuminates the profound impact of quantum computing on fields spanning science, technology, and society, promising to reshape our understanding of computation and accelerate the frontiers of human knowledge.

The Uncharted Future: Exploring Beyond the Horizons

As we stand at the crossroads of ever-evolving science and technology, the uncharted future beckons with the promise of groundbreaking discoveries, transformative innovations, and unprecedented possibilities. From the realms of artificial intelligence and space exploration to the frontiers of sustainable energy and biotechnology, this journey embarks on a speculative exploration of the uncharted territories that lie ahead. While the specifics remain uncertain, the trajectory of human progress is guided by the pursuit of knowledge, the spirit of inquiry, and the relentless quest to expand the boundaries of human understanding.

1. Artificial Intelligence and Machine Learning: Unveiling Intelligent Machines: The evolution of artificial intelligence and machine learning is poised to reshape industries, from healthcare and finance to entertainment and transportation, raising questions about ethics, automation, and human-machine collaboration.

2. Space Exploration and Colonization: Humanity's Cosmic Odyssey: The next era of space exploration may involve moon bases, Mars colonies, and even interstellar missions, transforming humanity into a multi-planetary species.

3. Climate Change Mitigation: Navigating the Path to Sustainability: The future demands innovative solutions to address climate change, including renewable energy technologies, carbon capture, and sustainable agriculture practices.

4. Biotechnology and Genetic Engineering: Rewriting Life's Code: Advances in biotechnology hold the promise of curing diseases, enhancing human capabilities, and reshaping the relationship between biology and technology.

5. Quantum Technologies: Harnessing Quantum Phenomena: Quantum technologies, from quantum computing to quantum communication, offer revolutionary capabilities for computation, cryptography, and secure information transfer.

6. Neurotechnology and Brain-Machine Interfaces: Bridging Mind and Machine: The fusion of neurology and technology could lead to direct brain-computer communication, enhancing human cognition and enabling new forms of interaction.

7. Synthetic Biology and Biofabrication: Designing Life and Materials: Synthetic biology may enable the creation of novel organisms and materials with tailored properties, revolutionizing industries from medicine to materials science.

8. Ethical Considerations and Societal Impact: Guiding Our Technological Future: As we forge ahead, addressing ethical concerns related to privacy, equality, and the responsible use of technology becomes paramount.

9. Unforeseen Discoveries and Paradigm Shifts: The Serendipity of Science: History has shown that many transformative discoveries emerged from unexpected avenues, reminding us that the future's true surprises lie beyond our current perceptions.

10. The Human Element: Collaboration and Imagination: In this uncharted future, the role of human creativity, curiosity, and collaboration remains central to shaping the course of progress.

11. Cultural and Philosophical Reflections: Navigating the Unknown: As we navigate uncharted frontiers, cultural, philosophical, and existential questions arise about our place in the cosmos and the implications of our innovations.

12. The Uncharted Continuum: An Ever-Evolving Journey: The uncharted future is not a destination but a continuum of exploration and discovery—an unfolding saga that extends beyond the limits of our imagination.

The uncharted future—a canvas of infinite possibilities—invites us to embark on a journey of innovation, exploration, and inquiry. While the specifics may remain shrouded in uncertainty, the relentless pursuit of knowledge and the human spirit of exploration propel us forward, enabling us to navigate the uncharted waters ahead, and continuing our legacy of transforming the world around us.

The Ongoing Journey: Exploring the Unseen and Uncharted

The ongoing journey of human exploration is a testament to our insatiable curiosity and relentless pursuit of understanding. From the depths of the ocean to the far reaches of space, we continue to unravel the mysteries of the unseen and uncharted realms that lie beyond our current knowledge. This expedition into the unknown encompasses the frontiers of science, technology, and human potential, revealing the interconnectedness of the universe and the boundless possibilities that await us.

1. Deep Ocean Exploration: Unveiling the Abyssal Depths: The ocean's depths remain one of the last great frontiers on Earth, holding secrets of ancient life, geological processes, and potential sources of renewable energy.

2. Microscopic Realms: The Hidden World of Microorganisms: Microscopic life forms, often invisible to the naked eye, play essential roles in ecosystems, medicine, and biotechnology, yet much of their complexity remains uncharted.

3. Mind and Consciousness: Navigating the Depths of Cognition: Understanding the human mind and consciousness is a profound journey that intertwines neuroscience, psychology, and philosophy, challenging our notions of self and perception.

4. Cosmic Mysteries: Probing the Nature of Dark Matter and Dark Energy: Dark matter and dark energy constitute the majority of the universe's content, yet their fundamental nature remains elusive, pushing the boundaries of astrophysics.

5. Multiverse and Quantum Realms: Beyond Our Known Reality: The theories of multiverse and quantum realms stretch the limits of our understanding, suggesting that reality might be more complex and interconnected than we can imagine.

6. Synthetic Reality and Augmented Humanity: Blurring Digital and Physical Worlds: Emerging technologies like virtual reality and augmented reality merge digital and physical realities, redefining communication, entertainment, and human experience.

7. Ethics and Morality in an Evolving Landscape: Guiding Ethical Choices: As we navigate uncharted territories, ethical considerations become crucial in shaping how we wield newfound knowledge and capabilities.

8. Cultural and Artistic Explorations: Expressing the Unseen: Art, literature, and culture offer unique lenses to explore the unseen and uncharted realms of human emotion, imagination, and expression.

9. Technological Evolution and Human Potential: Pushing Boundaries: Advancements in biotechnology, nanotechnology, and artificial intelligence hold transformative potential, challenging us to ponder the ethical implications of enhancing human capabilities.

10. Uncharted Journeys in Literature and Philosophy: Imagining the Beyond: Literature and philosophy take us on metaphorical journeys into the uncharted, illuminating the human condition and posing existential questions.

11. Spirit of Exploration and the Future Horizon: Beyond the Horizon Awaits: The spirit of exploration transcends generations, propelling us to venture beyond our current limits and ushering in an era of unparalleled discovery.

12. The Unseen and Uncharted: A Continuum of Discovery:

The uncharted is not an endpoint but an ongoing continuum of exploration, where the unseen remains an invitation to uncover the limitless wonders that the universe holds.

The ongoing journey of exploration unfolds as a tapestry woven with scientific inquiry, technological innovation, artistic expression, and philosophical contemplation. As we traverse the realms of the unseen and uncharted, we embark on a collective quest to illuminate the mysteries of existence, define the boundaries of our knowledge, and reach toward the stars with the unyielding spirit that defines us as curious, creative, and boundlessly ambitious beings.

Embracing Curiosity: Forging New Paths in Scientific Exploration

Curiosity is the driving force that propels humanity forward on a journey of scientific exploration, pushing the boundaries of knowledge and reshaping our understanding of the world around us. From the smallest particles to the vastness of the cosmos, curiosity compels us to ask questions, challenge assumptions, and seek answers that illuminate the mysteries of existence. This exploration celebrates the power of curiosity as a catalyst for discovery, innovation, and the relentless pursuit of truth that shapes the course of science and the trajectory of human progress.

1. The Curious Mind: Igniting the Spark of Discovery: Curiosity is the birthplace of scientific inquiry, inspiring us to ask questions, challenge assumptions, and embark on journeys of exploration.

2. Observing Nature: The Foundation of Scientific Inquiry: Observation is the starting point of scientific investigation, enabling us to gather data and uncover patterns that guide our understanding.

3. Questioning the Unknown: Unveiling Hidden Truths: Curiosity encourages us to question the unknown, to challenge conventional wisdom, and to seek answers that expand the boundaries of knowledge.

4. Hypotheses and Experiments: Testing the Limits of Understanding: Forming hypotheses and conducting experiments allow us to test our ideas, refine our understanding, and validate our insights.

5. Paradigm Shifts: Rewriting the Scientific Narrative: History is punctuated by paradigm shifts—transformative moments when new discoveries challenge prevailing theories and reshape the scientific landscape.

6. Interdisciplinary Exploration: Bridging the Gaps of Knowledge: Curiosity transcends disciplinary boundaries, fostering collaborations between diverse fields and sparking innovative solutions to complex challenges.

7. Ethical Dimensions of Curiosity: Navigating the Boundaries: While curiosity drives progress, ethical considerations ensure that scientific exploration respects human values, societal needs, and environmental impacts.

8. Curiosity-Driven Research: Fueling Discoveries and Innovations: Curiosity-driven research, driven by a passion to explore the unknown, often leads to unexpected breakthroughs and practical applications.

9. Curiosity in Space Exploration: Expanding Cosmic Frontiers: Space exploration embodies humanity's curiosity to explore the cosmos, uncovering new planets, phenomena, and insights into the origins of the universe.

10. The Next Curious Generation: Nurturing a Culture of Inquiry: Fostering curiosity in the next generation of scientists and thinkers ensures that the pursuit of knowledge continues to shape our future.

11. The Joy of Curiosity: Finding Wonder in Discovery: Curiosity brings a sense of wonder and awe to our encounters with the natural world, reminding us of the beauty and complexity that surrounds us.

12. Embracing Curiosity: A Lifelong Journey: The journey of scientific exploration is a lifelong pursuit—one that invites us to embrace curiosity, never stop asking questions, and continually push the boundaries of what we know.

Embracing curiosity is not just a call to exploration; it's an invitation to engage with the world with wonder, to be unafraid of the unknown, and to foster a spirit of inquiry that shapes the course of human understanding. As we navigate the uncharted territories of science, curiosity stands as our guiding light, propelling us to uncover the mysteries of the universe and to forge new paths in the pursuit of knowledge.

Key Terms and Concepts Defined

Here are definitions for some key terms and concepts discussed in the previous sections:

1. **Alkanes:** Saturated hydrocarbons composed of carbon and hydrogen atoms, connected by single covalent bonds.
2. **Alkenes:** Unsaturated hydrocarbons containing at least one carbon-carbon double bond.
3. **Alkynes:** Unsaturated hydrocarbons containing at least one carbon-carbon triple bond.
4. **Alcohols:** Organic compounds containing a hydroxyl (-OH) functional group attached to a carbon atom.
5. **Ethers:** Organic compounds characterized by an oxygen atom sandwiched between two carbon atoms.
6. **Aldehydes:** Organic compounds with a carbonyl group ($C=O$) at the end of a carbon chain.
7. **Carboxylic Acids:** Organic compounds featuring a carboxyl group (-COOH), found at the end of a carbon chain.
8. **Aromaticity:** A property of cyclic compounds with a conjugated pi electron system, exemplified by benzene.
9. **Heterocycles:** Cyclic compounds containing at least one atom other than carbon in the ring, like nitrogen, oxygen, or sulfur.
10. **Nucleophile:** An electron-rich species that donates electrons to form a new covalent bond.
11. **Electrophile:** An electron-deficient species that accepts electrons to form a new covalent bond.
12. **Reaction Mechanism:** A step-by-step description of how

a chemical reaction progresses at the molecular level.

13. **Substitution Reaction:** A reaction in which an atom or group in a molecule is replaced by another atom or group.

14. **Elimination Reaction:** A reaction in which a molecule loses atoms or groups to form a double bond or a ring.

15. **Addition Reaction:** A reaction in which two or more molecules combine to form a single product molecule.

16. **Electrophilic Aromatic Substitution:** A reaction in which an electrophile substitutes a hydrogen atom in an aromatic compound.

17. **Oxidation-Reduction (Redox) Reaction:** A reaction involving the transfer of electrons between reactants.

18. **Rearrangement Reaction:** A reaction in which atoms or groups within a molecule are rearranged to form a new isomer.

19. **Retrosynthetic Analysis:** The process of deconstructing a complex molecule into simpler fragments to plan its synthesis.

20. **Protecting Group:** A temporary modification used to block a reactive functional group during a synthesis.

21. **Spectroscopy:** The study of interactions between matter and electromagnetic radiation, providing information about molecular structure and properties.

22. **Quantum Mechanics:** A fundamental theory describing the behavior of particles at the atomic and subatomic scale.

23. **Transition State:** The highest energy point in a reaction pathway, where bonds are in the process of breaking and forming.

24. **Equilibrium Constant:** A measure of the ratio of reactants to products at equilibrium in a reversible reaction.

25. **Entropy:** A measure of the randomness or disorder of a system; often associated with the second law of thermodynamics.

26. **Curiosity:** The innate desire to explore, question, and understand the world around us.
27. **Qubits:** Quantum bits, the fundamental units of quantum information that leverage quantum superposition and entanglement.
28. **Paradigm Shift:** A fundamental change in the way a scientific discipline or society perceives and understands the world.
29. **Ethical Considerations:** Moral principles and values that guide the responsible conduct of scientific research and technological advancement.
30. **Synthetic Biology:** The engineering of biological systems to create new organisms or modify existing ones.

These definitions offer a glimpse into the rich tapestry of concepts and terminology that populate the world of organic chemistry, spectroscopy, quantum mechanics, and scientific exploration.

Cited Sources and Further Reading

Here are some cited sources and further reading materials related to the topics discussed:

Organic Chemistry:

1. "Organic Chemistry" by Jonathan Clayden, Nick Greeves, and Stuart Warren.
2. "Organic Chemistry" by Paula Yurkanis Bruice.
3. "Advanced Organic Chemistry" by Francis A. Carey and Richard J. Sundberg.

Spectroscopy:

1. "Introduction to Spectroscopy" by Donald L. Pavia, Gary M. Lampman, and George S. Kriz.
2. "Principles of Instrumental Analysis" by Douglas A. Skoog, F. James Holler, and Stanley R. Crouch.
3. "Modern Spectroscopy" by J. Michael Hollas.

Quantum Computing:

1. "Quantum Computing: A Gentle Introduction" by Eleanor Rieffel and Wolfgang Polak.
2. "Quantum Computation and Quantum Information" by Michael A. Nielsen and Isaac L. Chuang.

Uncharted Frontiers and Future Exploration:

1. "Packing for Mars: The Curious Science of Life in the Void" by Mary Roach.
2. "The Singularity Is Near: When Humans Transcend Biology" by Ray Kurzweil.

3. "The Future of Humanity: Terraforming Mars, Interstellar Travel, Immortality, and Our Destiny Beyond Earth" by Michio Kaku.

Curiosity and Scientific Exploration:

1. "The Curious Mind: How to Spark Your Curiosity, Increase Your Passion, and Empower Your Life" by Nir And Far.
2. "Curious: The Desire to Know and Why Your Future Depends on It" by Ian Leslie.
3. "A Short History of Nearly Everything" by Bill Bryson.

These resources offer in-depth insights and knowledge on the subjects covered and can provide a deeper understanding of the diverse topics discussed.

www.ingramcontent.com/pod-product-compliance
Lightning Source LLC
Chambersburg PA
CBHW012304240726

48656CB00008B/2542